Novel Antennas

Compiled and edited by
Steve Telenius-Lowe, PJ4DX

Radio Society of Great Britain

Published by the Radio Society of Great Britain
3 Abbey Court, Priory Business Park, Bedford MK44 3WH, UK
Tel: 01234 832700. Web: www.rsgb.org

Published 2015.

ISBN: 9781 9101 9310 5

Cover design: Kevin Williams, M6CYB

Design and layout: Steve Telenius-Lowe, PJ4DX

Production: Mark Allgar, M1MPA

Printed in Great Britain by Latimer Trend of Plymouth

Publisher's Note:
The opinions expressed in this book are those of the author(s) and are not necessarily those of the Radio Society of Great Britain. Whilst the information presented is believed to be correct, the publishers and their agents cannot accept responsibility for consequences arising from any inaccuracies or omissions.

Contents

Preface

WHEN I FIRST DISCUSSED the idea behind this book with RSGB Commercial Manager, Mark Allgar, M1MPA, my first question was "just what *is* a 'novel' antenna anyway?" Off the top of my head I could only think of the Crossed Field Antenna (CFA) and the E-H Antenna as falling into that category. But, looking out of my window in Bonaire I saw my own HF antenna: a Spiderbeam, providing the equivalent of a full-size 3-element monoband Yagi on 20m, a 2-element Yagi on 17m, a 3-element Yagi on 15m, a 2-element Yagi on 12m and a 4-element Yagi on 10m. Yes, a 14-element beam all on a single boom, fed with a single length of coaxial cable, containing no traps and light enough for one person to hold aloft. This, surely, fitted the description of a 'novel' antenna? I decided that, in addition to 'unusual' or controversial antennas, the term also included those where the designer had come up with an innovative take on an earlier or perhaps better-known design.

For that reason, this book does not include designs for standard Yagis, quads, quarter-wave verticals, horizontal loops, delta loops, magnetic loops, off-centre-fed dipoles, Windoms, Beverages, the Moxon Rectangle or any of the myriad antennas that are the standard fare of antenna books throughout the world. But although you won't find a standard dipole in this book, you *will* find the choke dipole; magnetic loops tuned by a variable inductance or a novel home-made capacitor rather than the usual vacuum variable; the 'Super Moxon', which adds a pair of directors to the standard Moxon Rectangle design; an orthoganally steered receive antenna that provides incredible levels of rejection of interfering signals; the home-made 'Wonder Whip' for QRP portable operation; a mobile antenna that can double as a car roof rack and, yes, the original Spiderbeam construction project described by its designer.

Novel commercially-made antennas are not excluded either, and independent reviews can be found of the unusual-looking Bilal Isotron antennas; the OptiBeam OBW10-5, which combines a five-element driver cell with both Moxon and Yagi-type elements in a novel hybrid design; InnovAntennas' new Opposing Phase Driven Element System ('OP-DES') and Loop Fed Array (LFA) designs, and several others.

Probably *the* most novel antenna in the book warrants a chapter of its own. In 2011 *RadCom* carried a four-part series, the 'PICaYAGI' by Peter Rhodes, G3XJP. This is a major constructional project (and not for the faint-hearted) in which the physical length of the elements are automatically adjusted to provide not just mono-band but mono-*frequency* performance. The PICaYAGI design is printed here in its entirety and as a single feature for the first time.

Thanks go to all the authors who have devoted time and effort to develop their antennas and then taken the trouble to write them up for publication so that we may all benefit from their work. Each of these authors is acknowledged in the text.

The ingenuity of radio amateurs around the world knows no bounds and that is reflected in the pages of this book. With designs from the USA, Canada, France, Germany, Italy, Sweden and even El Salvador – as well as the UK of course – we feel sure there will be something of interest to all antenna experimenters.

Steve Telenius-Lowe, PJ4DX
Bonaire, March 2015

Chapter 1
Dipoles

THE INVERTED-U ANTENNA

You've heard of the inverted-V antenna and probably also the inverted-L. Well, this is the inverted-U. L B Cebik, W4RNL, described this antenna in the May 2005 *QST* as a simple, small and light-weight design for use on field day operations and the like.

A dipole's highest current occurs within the first half of the distance from the feedpoint to the outer tips. We lose very little performance if we fold, bend or otherwise mutilate the outer end sections to fit an available space. If we start with a tubular 10m dipole – 16ft 8in (5.08m) in overall length – we might add extensions for 12, 15, 17 or 20m, shaping them to suit the area in which we are operating. If we can find space to erect a 10m rotatable dipole at least 20 feet (6m) above the ground, with a clear area to permit us to rotate the dipole, then we can simply let the extensions hang down. **Fig 1.1** shows the relative proportions of the antenna on all bands from 10 to 20m. The 20m extensions are the length of half the 10m dipole. Safety dictates an antenna height of at least 20ft (6m) to keep the tips above the 10ft (3m) level. At any power level, the ends of a dipole are at a high RF voltage while transmitting and we must keep them out of

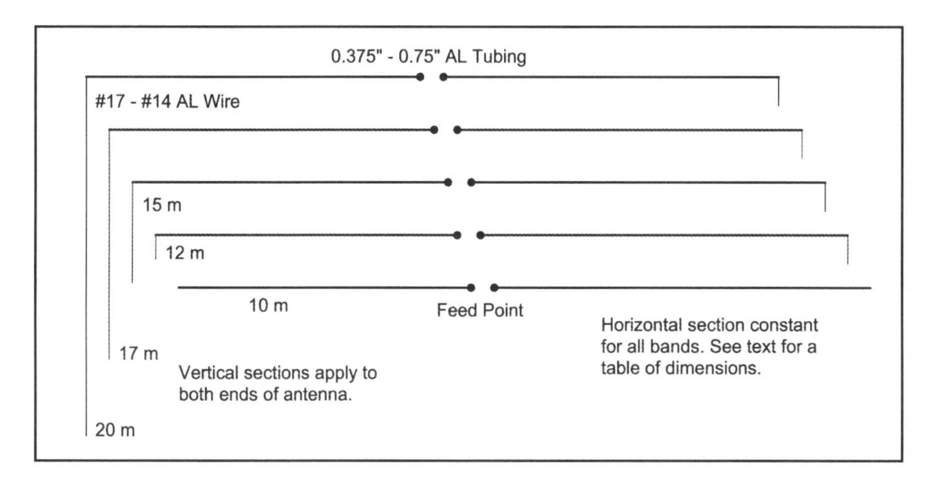

Fig 1.1: The general outline of the inverted-U field dipole for 20 to 10m. Note that the vertical end extension wires apply to both ends of the main 10m dipole.

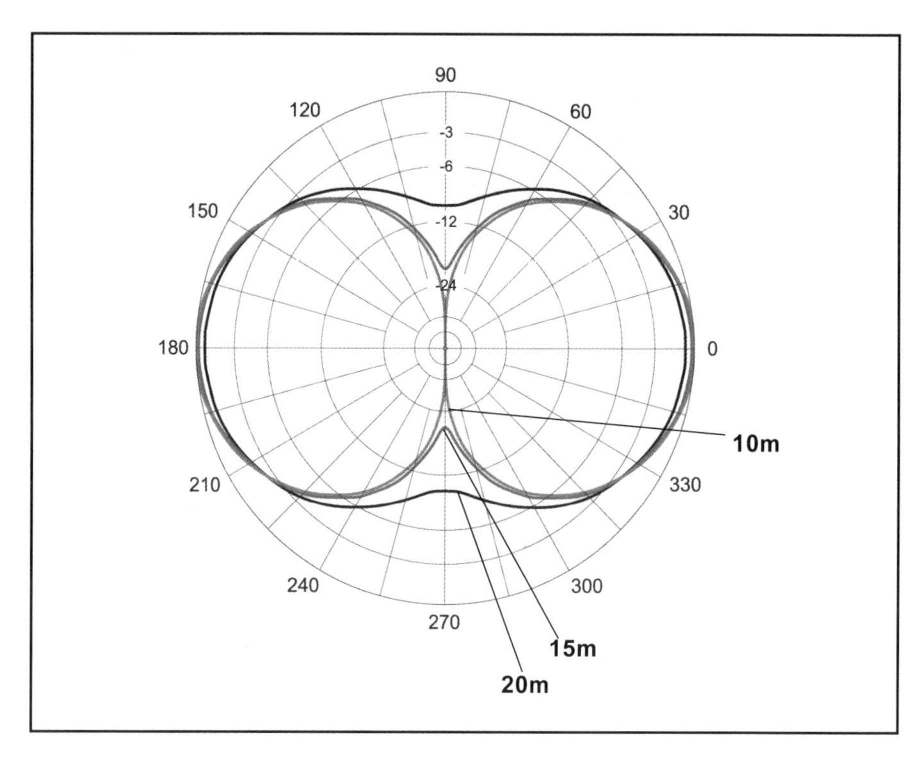

Fig 1.2: Free-space E-plane (azimuth) patterns of the inverted-U for 10, 15 and 20m, showing the pattern changes with increasingly longer vertical end sections.

contact with humans or animals.

In principle, we do not lose very much signal strength by drooping up to half the overall element length straight down. What we lose in bidirectional gain shows up in decreased side-nulls as we increase the length of the drooping section. **Fig 1.2** shows the free-space E-plane (azimuth) patterns of the inverted-U with a 10m horizontal section. There is an undetectable decrease in gain between the 10m and 15m versions. The 20m version shows a little over a half dB broadside gain decrease and a signal increase off the antenna ends. On 20m, the current in the vertical wires becomes significant, rounding the pattern.

The real limitation of an inverted-U is a function of the height of the antenna above ground. With the feedpoint at 20ft above ground, we obtain the elevation patterns shown in **Fig 1.3**. The 10m pattern is typical for a dipole that is about 5/8-λ above ground. On 15m, the antenna is only 0.45λ high, with a resulting increase in the overall elevation angle of the signal and a reduction in gain. At 20m, the angle grows still higher, and the signal strength diminishes as the antenna height drops to under 0.3λ. Nevertheless, the signal is certainly usable. A full-size dipole at 20m would show only a little more gain at the same height, and the elevation angle would be similar to that of the inverted-U, despite the difference in antenna shape. As we decrease our frequency, there is no substitute for antenna height. Any horizontal antenna below about 3/8-λ in height will show a rapid decrease in low angle performance relative to heights above that level. If we raise the inverted-U to 40ft, the 20m performance would be very similar to that shown by the 10m elevation plot in **Fig 1.3**.

Table 1.1 summarises the free-space and 20ft (6m) performance characteristics of the inverted-U. Of special note is the fact that the feedpoint impedance of the inverted-U remains well within acceptable limits for virtually all equipment, even at

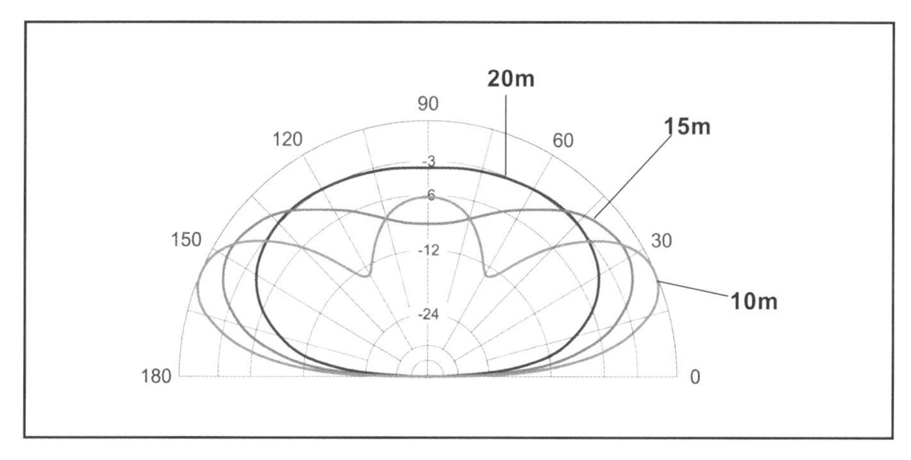

Fig 1.3: Elevation patterns of the inverted-U for 10, 15 and 20m, with the antenna feed point 20ft (6m) above average ground. Much of the decreased gain and higher elevation angle of the pattern at the lowest frequencies is due to its lower height as a fraction of a wavelength.

a height as low as 20ft above ground. Finding exact dimensions, even in field day conditions, becomes a non-critical task. If we can accept the performance potential of a dipole for any band in the 20m to 10m range at the anticipated height of our mast possibilities, then the inverted-U provides a compact way of achieving that performance.

The May 2005 *QST* article by L B Cebik includes full constructional details.

	Free-Space			20ft Above Average Ground		
Band (m.)	Gain (dBi)	Resonant impedance (ohms)	Wire length (cm)	Gain (dBi)	Elevation Angle (deg)	Impedance R ± *j*X (ohms)
10	2.1	73	—	7.6	24	65 – *j*2
12	2.0	71	41	7.2	27	67 – *j*8
15	1.9	64	97	6.4	32	69 – *j*8
17	1.7	55	157	5.7	38	65 – *j*4
20	1.4	41	274	4.8	50	52 + *j*4

Using a tubular 10m dipole and AWG17 to 14 vertical wire element extensions. The wire length for the drooping ends is measured from the end of the tubular dipole to the tip of the wire. Little change in length occurs as a function of a change in wire size. However, attachment of the extension to the element and special field conditions may require a few extra inches of wire.

Table 1.1: Anticipated performance of the inverted-U for 20 to 10m.

THE 'VISTA' – A SMALL ANTENNA FOR QRP USE

'VISTA' in this case stands for 'Variable Inductance Small Telescopic Antenna' and is a small dipole intended for QRP operation (only). Its designer, Dr John Seager, G0UCP, describing it in the June 2010 *RadCom*, likened the fun of experimenting with small antennas with that of low-power (QRP) communications: it is all about achieving more with less. In the case of small antennas, if they can be easily

The complete VISTA antenna, shown with the two 1.3m long 10 section telescopic dipole elements retracted. Extended, the antenna is just over 2.6m wide.

carried one can accept a degree of inefficiency in exchange for convenience of use.

The VISTA fits into that category. It is a centre-loaded shortened dipole but because it uses a ferrite rod it is suitable only as a low-power device. But while it is not an efficient radiator, it is very handy, needs no pole or guys to keep it up and requires no separate matching unit. It also needs no counterpoise or earth connection, making it suitable for portable QRP operations from locations anywhere.

John Seager, G0UCP, describes how he came up with the design: "The idea came from a simple antenna matcher ['No Cost ATU', by Tony Haas, G4LDY, in *Sprat* 28, Autumn 1987 – *Ed*] that used a ferrite rod to vary the inductance of a self-supporting coil. It occurred to me that something similar might be useful to resonate a small dipole. Initially I took two telescopic lecture pointers (each about 60cm long) and wound a loading coil at the centre. The coil was loop coupled via a length of RG174 coaxial cable to a bidirectional power meter and a SoftRock / Power SDR transceiver tuned to the 14MHz band. As the ferrite rod brought the system to resonance there was a gratifying increase in the basal noise trace and several CW signals could be tuned in on the receiver. Even more encouragingly, with a series capacitance of about 80pF included in a central coupling link, it was possible to achieve a 1:1 match to the 50Ω transceiver output. With minor changes, similar results were obtained on each band from 7 to 28MHz. Still with the short 'pointer'

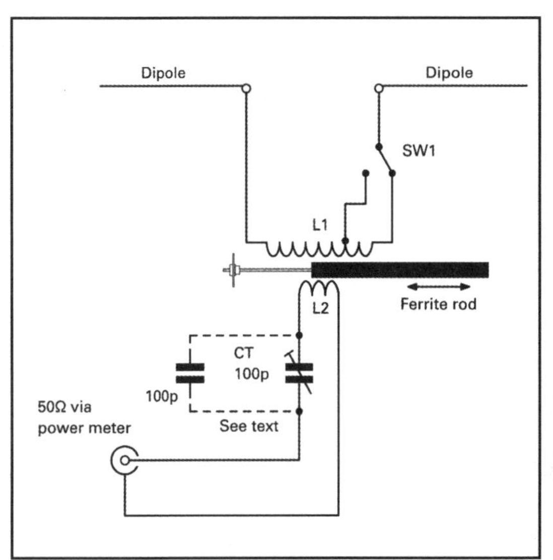

Fig 1.4: Circuit diagram of the VISTA. See text and the Table for component and coil winding details.

	LF version	HF version
CT (nominal)	80pF	60pf
Additional capacitor	100pF	none
L1	10 turns	14 turns
Tap L1 at	10 turns	4 turns
L2	2 turns	2 turns

Table 1.2: Component values for lower and higher-frequency versions.

elements in place and with the antenna in a ground floor room, I heard a strong CQ call on 18MHz from a station near Kiev, Ukraine. No one answered, so I called him back. He copied my callsign and gave a 339 report. Output power was 2W and the antenna was just 1.3m above floor level."

The circuit of the VISTA antenna is quite simple and is shown in **Fig 1.4**, while **Table 1.2** gives component values for the low- and high-band versions. The central components are L1 and L2 that are coupled to a varying degree by the ferrite rod. Variable capacitor CT provides matching to the incoming 50Ω line (supplemented by an additional 100pF for the lower frequency version). L1 directly drives the telescopic elements of the antenna, with a couple of taps at one end selected by switch S1.

The main difference between the low- and high-frequency versions is the coil L1. The coil for higher frequencies has 14 turns tapped 4 from the end, wound on a 2.5cm plastic tube. The lower frequency coil had 30 turns, tapped 10 from the end. The series capacitor CT can be a pre-set trimmer or a small variable (e.g. Maplin FT78K) mounted inside the box. In the lower frequency version it may be beneficial to add a 100pF capacitor in parallel with the trimmer for the lower frequency bands (see **Table 1.2**).

In a later development of the high frequency version G0UCP used thicker wire, oval in cross section and equivalent to about 16SWG. Twenty turns resonated at 14, 18, 21 and 24MHz. If the coupling winding is placed a few mm to the right of the loading coil (i.e. the end first entered by the ferrite rod on its insertion) it is possible to achieve a 1:1 match with no series capacitor.

A further refinement, leading to only a small increase in size, was to build the unit into a 'Double Mounting Box' with the coil and tuning mechanism in the lower compartment. This had the advantage that there was negligible hand capacitance effect when making adjustments to the tuning.

In place of the original pointers, Maplin 10 section telescopic (1.3m) antennas (code LB10) were used. These have a hole in the base tapped to take an M4 bolt. Most of the hardware came from B&Q and Homebase. Everything is housed in a 'Single Mounting Box' intended for domestic electrical use and made from synthetic plastic 25 or 30mm deep. With suitable precautions it can be hand drilled, and might have been expressly designed for the VISTA antenna.

To move the inductor core, a 40mm long 3.5mm diameter machine screw was threaded through a cup formed from a piece of graduated rubberised cable protector. The screw thread then passes through a nut secured with cyanoacrylate adhesive (or, better still, a stripped-down and slightly crimped stereo jack socket) in the case wall, so that the ferrite rod is held firmly. Its other end protrudes through a hole drilled in the opposite side of the box as shown in the photo to the right. The rod can be moved in and out by rotating the screw or the rod itself, allowing precise and stable tuning.

The VISTA will tune to give a 1:1 50Ω match on any of the HF bands in its range. G0UCP uses 4m of RG174 to connect it to the power meter and transceiver. In use, the antenna is set up on a suitable support and peaked for maximum receiver noise level. It is then adjusted for zero reverse power on transmit. The series capacitor can be pre-set to about 80–100pF for 14MHz, a bit more for 7MHz and much less for the higher bands. It need only be adjusted once for each band. Antenna resonance is affected by proximity to the operator, so it is necessary to tune in increments and check several times. If peaked for the middle of the 20m CW section it is possible to operate from 14.000 to 14.060MHz without retuning.

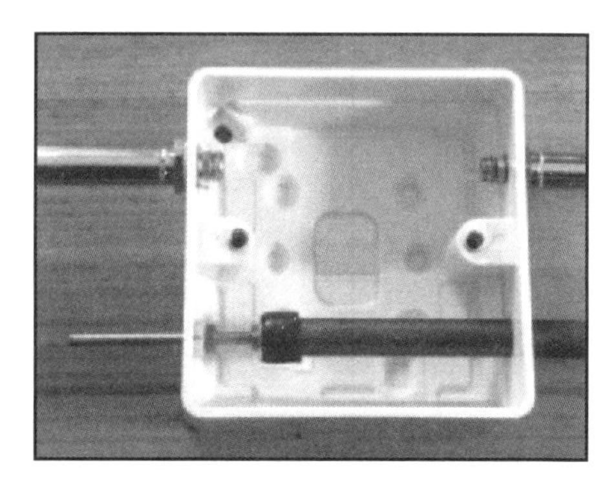

Mounting arrangements for the ferrite rod and dipole elements.

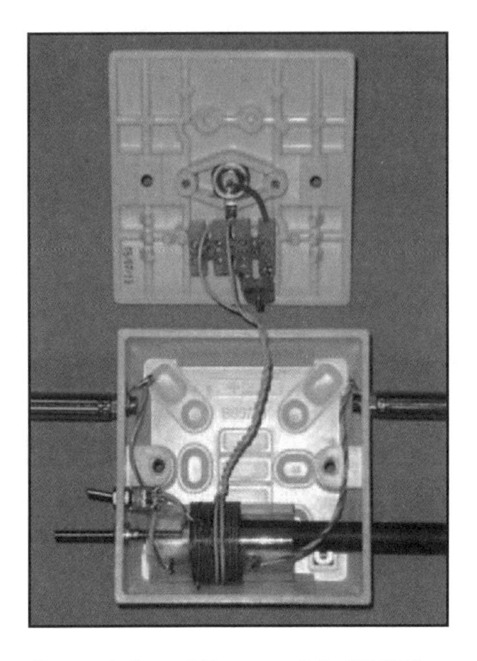

General view of the completed VISTA.

Alterations to the length of the coax feed made no difference to antenna resonance. A 4m length was enough to site the antenna clear of obstructions but within easy reach to make adjustments.

Remember, this is a *QRP-only* antenna. Do *not* attempt to drive it with more than 5W.

With the VISTA in a downstairs room G0UCP had six brief CW contest exchanges with east coast USA and Canadian stations. One of these was with 2W and the others between 3.5 and 5W. Two more US contacts were made with the antenna sited in the roof space and 4.5W. The increased height and 10–12m of extra feed line did not seem to make any difference to contacts to the USA. Aside from contests, he has made several contacts around Europe lasting for five minutes or more.

To conclude, this is a small antenna that works. It is not an efficient radiator, though this would be improved if its electrical length were increased, perhaps with capacity hats at the ends. You do need to check the reflected power during operation. It is quick and unobtrusive to set up and it fits not just in a briefcase, but in a sponge bag if it is dismantled. As it works indoors it might do even better on top of a small mountain!

A CHOKE DIPOLE

The sleeve dipole concept is familiar enough territory, though it seems few people actually build one, for reasons we will discuss in a moment. But a sleeve dipole *without* the sleeve – well, that *is* novel! This variation on the sleeve dipole by Robert Zimmerman, VE3RKZ, was described "for antenna experimenters to try" in the July / August 2010 edition of the ARRL publication *QEX*. The choke dipole allows the antenna to be end fed with 50Ω coax, making it convenient for city flat dwellers, or anyone who may prefer an end-fed antenna arrangement.

Fig 1.5 shows the standard sleeve dipole, with the coaxial feed line, going to the centre of the dipole, routed through a quarter-wave conducting sleeve. The dipole sinusoidal RF current distribution is impressed on the sleeve and a quarter-wave wire forming the second half of the dipole. This geometry permits convenient feeding of the dipole from one end. Note that the electric field, E, at the ends of a dipole are maximum, so care must be taken with this high voltage feed point. Despite the

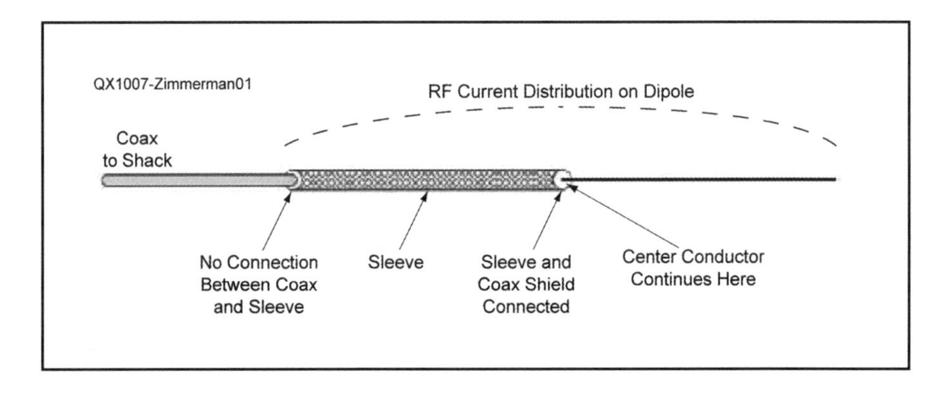

Fig 1.5: Construction of a sleeve dipole.

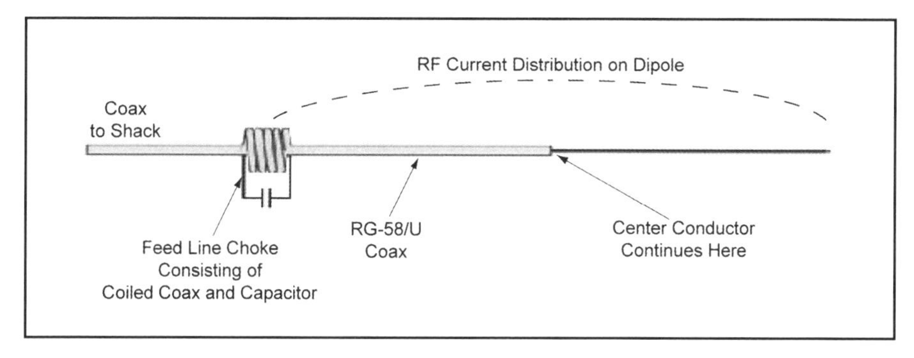

Fig 1.6: Construction of the choke dipole.

convenient geometry, a sleeve dipole is not all that convenient to construct, since it entails the tedious job of dressing a copper braid over a long piece of coax. This article presents a 'sleeve dipole equivalent antenna' – but minus the sleeve.

The feed end of the sleeve forms one end of the dipole RF current distribution. Radio frequency current cannot bridge the sleeve / coax gap, as the gap represents a very high RF impedance. It is proposed to replace the sleeve and high impedance gap with a high impedance current choke, as shown in **Fig 1.6**. The sinusoidal dipole current then may be impressed on the exterior surface of the coaxial shield and the same quarter-wave wire as before.

To make an effective current choke, VE3RKZ wound the coaxial cable to form a coil, and bridged it with an air variable capacitor to form a high impedance parallel LC network. For such a choke, we need to have some idea of the effective inductance and capacitance of the dipole itself. That is to say, we need to know the equivalent circuit model of a resonant dipole antenna. For this equivalent circuit, VE3RKZ turned to *EZNEC* and modelled a half-wave dipole in free space for 14.200MHz. He used a 1/4in wire diameter to represent RG-58/U coax. A length of 10.181m provides the resistance and reactance values shown in **Fig 1.7**. If this impedance,

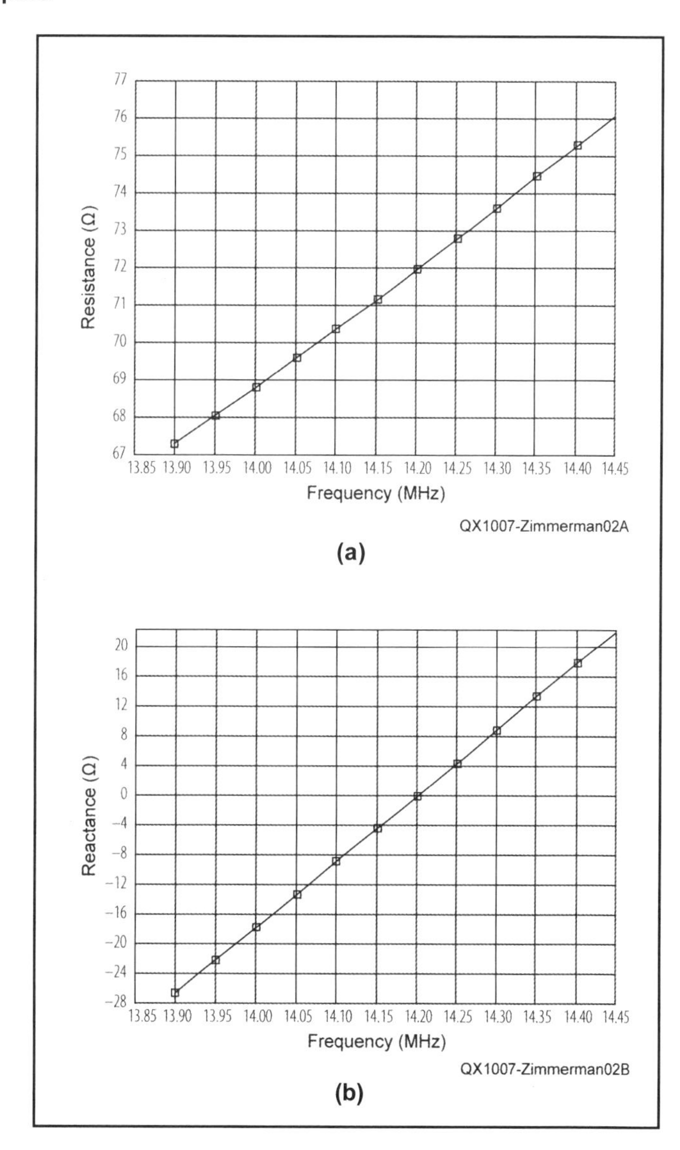

Fig 1.7: (a) Resistance of 14.200MHz dipole, as modelled on *EZNEC*. (b) The modelled reactance of the same.

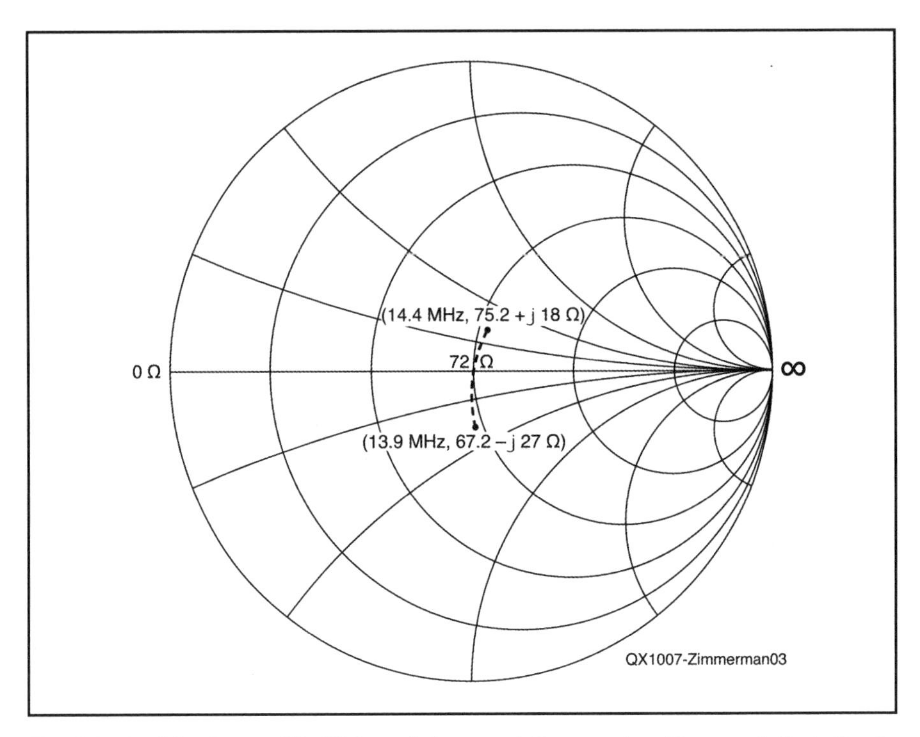

Fig 1.8: Smith chart with 72Ω resistance at the centre. The modelled dipole impedance is slightly shifted from the 72Ω circle.

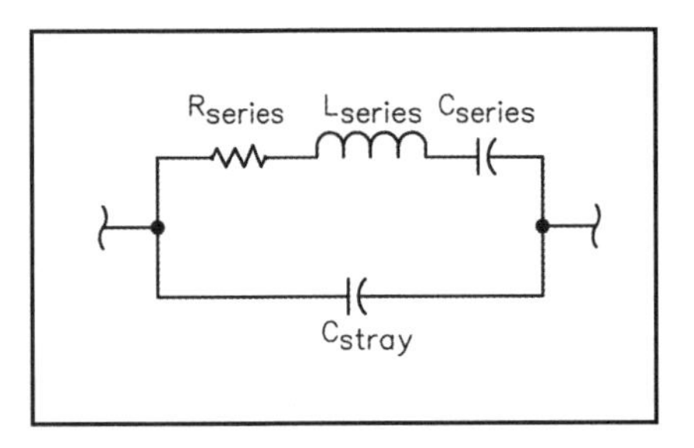

Fig 1.9: This is the proposed circuit model of a resonant dipole antenna.

The 1000Ω impedance current choke for 14.200MHz.

Z = R +*j*X, is plotted on a Smith Chart (**Fig 1.8**), we see the locus is very nearly the real 72Ω circle. Well, not quite: the entire circle is rotated clockwise around the origin, as if bridged by a stray capacitance. Accordingly, the candidate circuit model shown in **Fig 1.9** was tested with the impedance data in **Fig 1.7(a)** and **Fig 1.7(b)**.

The proposed circuit model of a resonant dipole antenna is given in **Fig 1.9**. The resulting component values for 14.200MHz are: R_{series} = 71.4062Ω, L_{series} = 7.2081µH, C_{series} = 17.5906pF, C_{stray} = 14.0387pF. These values they reproduce the dipole impedance exactly (to within 0.01Ω for the antenna R and X) between 13.9MHz and 14.4MHz. Those interested in using this model for another band at frequency *f* (MHz) may scale the inductor and two capacitors by the ratio 14.200 / *f* (MHz). There is no need to scale the resistance, R_{series} as it remains the same independent of the dipole frequency.

A good starting point for choke design is to use inductor and capacitor values approximately equal to those in the dipole itself. For the 20m inductor, a target value of 7µH was chosen. The resonating capacitor for 7µH and 14.200MHz is 24pF. Using the *ARRL Single-Layer Coil Winding Calculator*, 10 turns of coax on a 3in diameter mandrel with a total coil length of 2.0in fitted the bill. After winding the coil of coax, a 3/4in segment of coax braid on each end of the coil was bared. The coil was bridged with a 10–100pF air variable capacitor, as shown

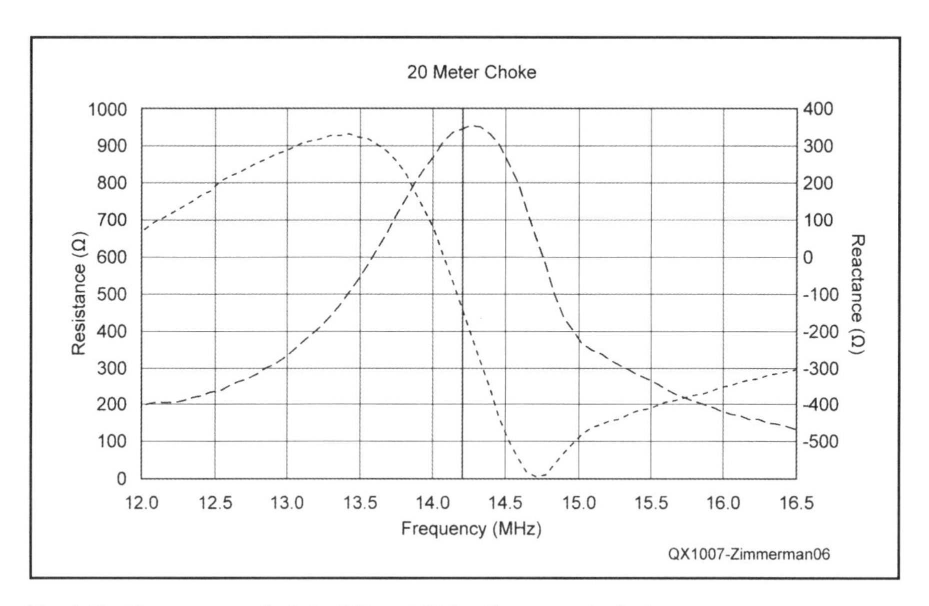

Fig 1.10: The measured plot of R and X for the current choke.

in the photo. The coil form was a plastic drink bottle: you can see strips of duct tape holding the coax coil turns in place.

To be sure the choke had sufficient impedance to keep RF current off the feed line, the choke was tested with an AEA Bravo VIA impedance analyser, using 1in test leads. The resulting series impedance of the parallel tank is 1000Ω, which was deemed satisfactory for 100W operation. The data is shown in **Fig 1.10**.

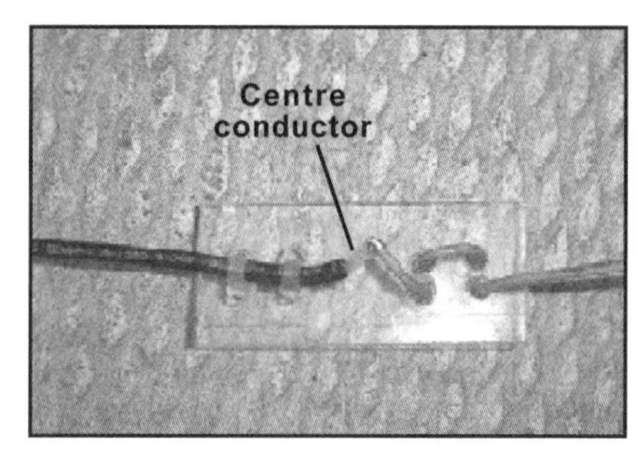

The centre feed point of the VE3RKZ Choke Dipole.

The total length for the resonant dipole, from *EZNEC*, is 10.181meters, half of which is RG-58/U coax and half of which is AWG12 copper wire. The dipole centre connection of the coaxial cable centre conductor to the AWG12 wire is shown in the photo.

VE3RKZ's back garden is small, and he had to install the dipole in a sloping manner, as shown in the photo. The choke is on top of the bamboo pole at the left, and the AWG12 wire is supported at the right on a 9m push-up mast.

The dipole initially resonated at 14.350MHz with an SWR of 1.46:1. The passband is primarily dependent on the dipole length, and less so on the choke tuning. The choke was

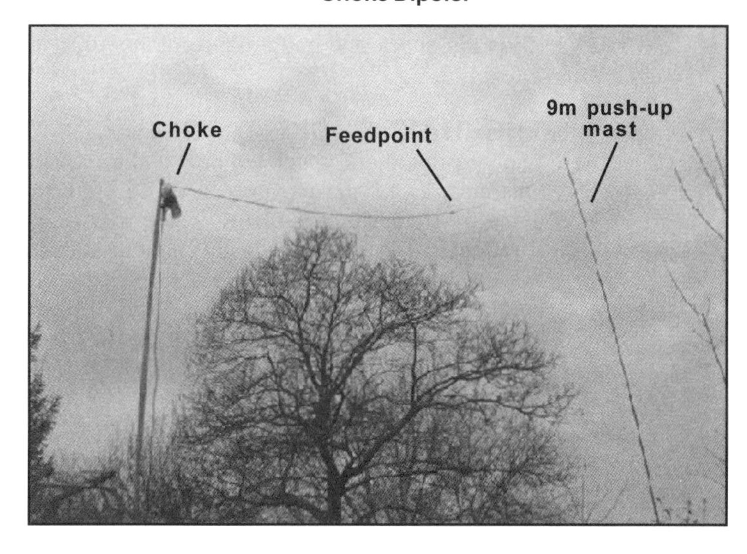

The sloping Choke Dipole in VE3RKZ's back garden.

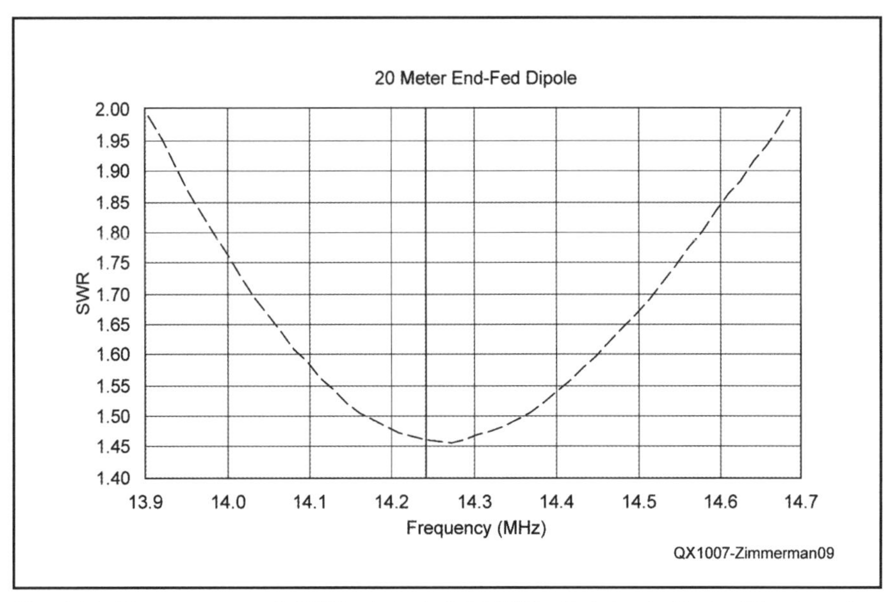

Fig 1.11: The measured SWR plot of VE3RKZ's finished Choke Dipole, after adjustment.

lowered briefly, and the capacitance slightly increased; this brought the antenna centre frequency to 14.250MHz as shown in **Fig 1.11**.

After tune-up, the choke (or at least the variable capacitor) should be protected in a weather-proof plastic box; do not use a metal box, as there is high RF voltage across the choke. You may want to measure the capacitance value after the antenna is adjusted for the best operation, and replace the variable capacitor with a fixed-value unit. You could also just build the antenna using a fixed value capacitor in the range of about 25 to 35pF. Be sure to choose one that can handle the high voltage point at the end of the antenna.

Initial tests were made at 25W PEP. W4YKY in southern Georgia came back with a 59+10dB report and said, "Don't change anything!" There is no evidence of RF in the shack and performance at the 100W level has been pleasing.

THE 'HAK' CHOKE DIPOLE

Showing that there is nothing new under the sun – even when it comes to 'novel' antennas – after the previous 2010 *QEX* article by VE3RKZ comes the following piece by Peter Gant, G8HAK, which was originally published in the April 2013 *RadCom*. G8HAK independently came up with a similar Choke Dipole idea and the two articles are published here one after the other for readers to make their own comparisons.

G8HAK wrote that some years ago a practice started to appear in antenna designs of coiling the coaxial feeder to form an RF choke at the dipole termination. This was to prevent RF flowing back along the surface of the feeder and thus affecting the normal characteristics of the dipole. The examples he came across seemed to him to be very hit and miss and unspecific, though. How could they be fully effective in preventing RF current from flowing over the feeder outer? To be fully effective the inductance formed by coiling the coax feeder would need to be quite precise at the frequency of interest.

It was intriguing that the coaxial choke balun / trap formed would be effective, yet

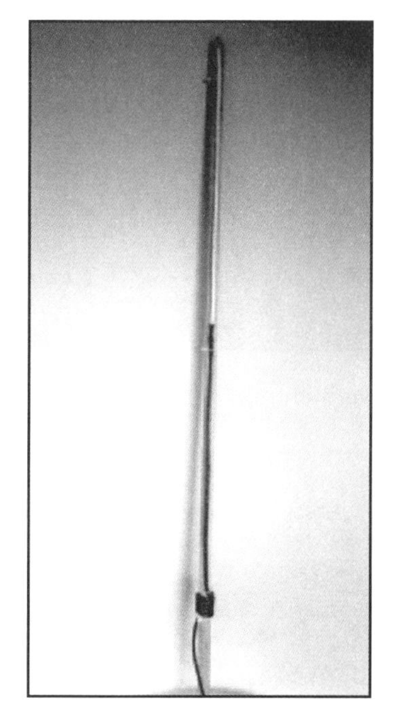

The 2m HAK Choke Dipole.

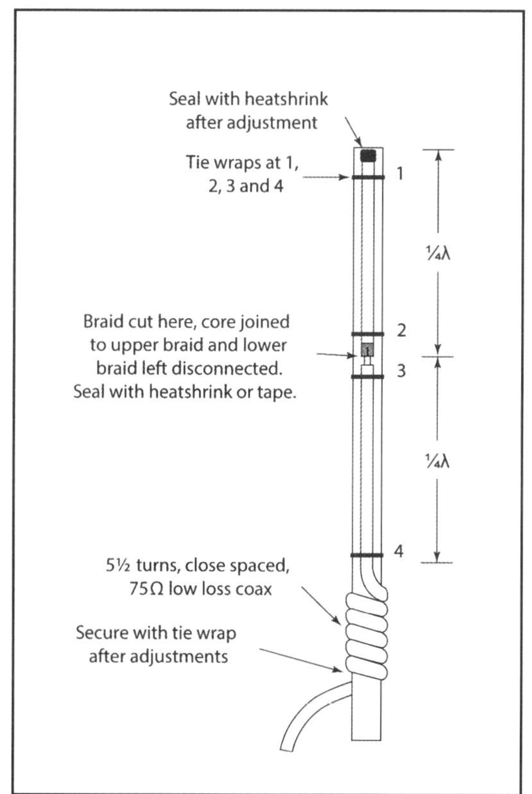

Fig 1.12: Basic construction of the HAK dipole (2m vertical version: compare with photo).

at the same time would not affect the normal function of the coax. If this was the case, could standard coax be employed to provide both the feeder and quarter-wave element of a classic dipole with the use of a coaxial choke?

G8HAK built a 2m example to provide a quick check of the theory. The prototype is shown in the photo above and in **Fig 1.12**. Once the element is prepared as shown, loosely secure the coax using tie-wraps at positions 1 – 4 and use a small amount of insulating tape to hold the bottom of the coil in place. With the antenna vertical and away from obstructions, check the VSWR using an antenna analyser. Trim the top section above tie 1 and adjust the bottom of the first turn of the coil until the best match is achieved. On the prototype, the final length between the tip of the element and the start of the coil section was 495mm (19.5in). The VSWR was 1.4:1 at 144MHz, 1.3:1 at 145MHz and 1.2:1 just below 146MHz.

This vertical was installed at gutter height and compared over some months with the main station Slim Jim, which is mounted some 5m higher. Given the height advantage of the main antenna the performance of the HAK Choke Dipole was about as expected from a standard dipole.

However, the higher HF bands were the main area of interest. The physical layout would have a number of advantages in that application, for both vertical and horizontal use. The next two examples were constructed for 10m and 17m. In each case a glass fibre fishing pole was used as the support, which was mounted on the top of a 3m aluminium pole secured to a greenhouse. The greenhouse is about 15m from the house and the antennas under test were fed by about 20m of buried RG-213.

The 10m vertical antenna mounted on a pole secured to the greenhouse.

Coil for 17m vertical version.

Coil for 20m horizontal version.

In the 2m antenna the outer braid was severed at the centre. The severed section was joined to the inner to form the second ¼ wave section. An option would be to remove sleeve and braid, leaving just the centre conductor and insulation for the second section. This was in fact done on the 10m version. On longer lengths we can remove the coax completely at the centre and extend the centre core with a plain wire of choice. The 20m horizontal example was constructed in that way. Any concerns about tensions on the joint and centre conductor were unfounded although any longer lengths would be in need of support: 50lb monofilament fishing line or similar would be suitable.

Given that dimensions for dipole elements are well documented and relevant formula readily available, that leaves us with constructing the chokes required for each antenna. 50mm (2in) formers were used, or the base of the glass fibre mast. 75Ω low loss TV coax was used in all cases as this resulted in the best compromise in terms of SWR, probably because it best suited the natural dipole impedance. A 75Ω to 50Ω transformer to match the antenna to the feeder. **Table 1.3** lists the coil data for 10, 17 and 20m antennas. The coil for the 17m version is shown in the photo at the top of the page.

Band	Turns
10m	9
17m	14, in 2 layers
20m	24, in 2 layers

Table 1.3: Coil data (see text).

This data is intended as a starting guide. As implied earlier, the final figure will require adjustments to fractions of a turn to obtain the best SWR and this is where an antenna analyser makes life easy. Added to this process is the need to have the antenna in its operating position to account for the effect of nearby objects.

Horizontal versions of this antenna can have some advantages, for example if your shack is at one end of the antenna layout, real benefits can be obtained: a shorter feeder can be used and the weight of choke and feeder is nearer to the end (reducing the sag). The antenna may also have a lower visual profile.

The choke for the 20m version, which is intended for horizontal use, is shown in the photo above. The coil former is part of a discarded sealant gun canister of about 50mm diameter.

Having used the 17m vertical for some weeks to good effect, G8HAK also made a horizontal version for a side-by-side comparison on receiving. The results were not as expected. He lives in a typical 1930s three-bedroom house on an estate, so expects the usual interference issues. What surprised him, however, was that the quieter background noise level on the vertical, often by as much as 1 to 2 S-points. Received signals were also stronger in about 75% of cases, even sometimes when the direction of transmission was broadside to the horizontal antenna. The horizontal antenna was fed directly from the upper floor shack window, passing over the greenhouse at a height of about 8m. The 17m vertical centre height was about a half-wave above ground level.

A NOVEL 5MHZ NVIS ANTENNA

Chris Saunders, G4ZCS, described this antenna in the March 2012 *RadCom*. He described how his local emergency services proposed a major railway evacuation exercise deep in the valleys of the South Downs, to which his radio club was invited to test communications links. VHF was unsuitable because of the terrain, so this prompted him to experiment with an easy to erect emergency antenna using HF NVIS. The exercise was eventually cancelled at the last minute by the police on health and safety grounds, but by then G4ZCS had developed the antenna.

It is a dipole supported on animal electric fence supports that are made largely of plastic but with a metal spike at the base. The theoretical height of this antenna should be about 0.25 of a wavelength above ground (14m); however, a practical NVIS antenna can be lower. A height of just a metre or so is ideal to make the antenna truly portable in a small car, easy to erect single handed in all weathers – and in the dark.

The 'T' piece at the centre of the dipole is a piece of engineering plastic from the Bluebell Railway track scrap bin. It was tested in the microwave for a few seconds

Feed arrangement. The tobacco tin contains a 1:1 / 4:1 balun.

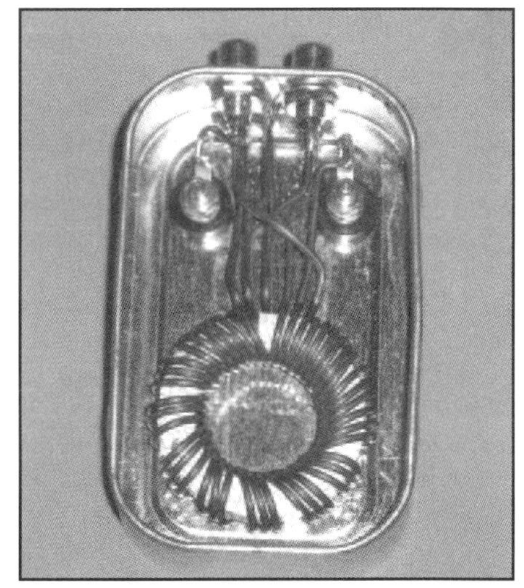

Detailed view of the balun (see text).

The air-cored balun, feed and centre support arrangement used by G3WYN.

alongside a cup of water: the plastic did not get warm, so it was concluded it was OK for RF. The balun is mounted in a 1oz tobacco tin. It is a dual ratio 1:1 – 4:1 transformer type toroid wound on a 40mm (1.5in) diameter ferrite ring. There are 14 trifilar turns of enamelled copper wire, separated from the ferrite by a layer of PTFE

tape. The unbalanced terminations are bought out of the enclosure using phono sockets – a phono to SO239 adaptor is used to couple the feeder. G3WYN used an air cored balun for a similar antenna, but any type of balun would suffice.

Using normal free space calculations, the length of a 5MHz half wave dipole antenna should be in the region of 28m. However, the proximity to the ground caused resonance to occur at a much shortened element length. The elements used by G4ZCS are each 9.13m from the feed point insulator, plus 10cm tails to the balun. The tails are 10cm apart on the insulator, making the total wire elements 9.23m; the antenna is 18.36 metres overall when laid out straight.

Using the various clip positions on the uprights the antenna was tested from 23cm (9in) up to 84cm (2ft 9in) above ground. The optimum seems to be at around 23 – 30cm (9 to 12in), giving a resonance centred at 5.35MHz and an impedance of just over 35Ω.

Initially, the radio picked up a lot of noise, but by connecting an earthing spike (a red-handled screwdriver) to the rig and keeping the clay ground wet, the noise level was reduced considerably.

Tests were carried out at two locations on three occasions. The first test was at Jack and Jill Windmills on the South Downs at 210m (700ft) ASL. The ground conditions were those of summer dry chalk. The antenna was run out in a straight line about 30cm above the ground, running roughly north-south. The whole station took less than five minutes to unpack from the car, assemble and tune up. During

General view of the antenna and portable operating position.

Operation from the quarry site. The antenna is close to the trees; the top of the cliff can be seen through their upper branches.

daylight hours contacts were made locally and as far away as East Anglia. After the sun had set more contacts were made, this time with the far south-west, from mid-Devon down to Cornwall and, in the other direction, into Essex. Most of these later contacts produced reports of 59+20dB, with G4ZCS using transmit powers in the range of 25 to 50W.

The following June, he operated again from the same site with similar ground conditions. Once again good signal reports were received from all over England and as far away as North Wales.

The final test was in October 2011 and on this occasion was from a deep valley site in an abandoned chalk quarry deep in the South Downs where there was no line-of-sight path available. It is horseshoe shaped and well over 30m (100ft) deep, The station was set up as before, with stakes in the soft turf, but they would not go in very deep because of the solid chalk beneath. There were tree canopies above, and the cliff face was about 50ft away. On that day there was a SFI of 147, the A index was 4 and the K index 2. On turning on the radio, some signals could be heard; after a quick tune up, G3WYN (12 miles away) was heard at 58. A call was made, and he gave G4ZCS/P, using 50W, a 55 report. G0WGP (7 miles away) reported 53 with high background noise, while G7TMR (9 miles) gave a 57 report.

NVIS was shown to be the transmission mode; no VHF path was available even to the local repeater network and only a weak mobile phone signal was obtainable at the quarry entrance (presumably from the nearby town of Lewes).

A low strung antenna wire is an obvious trip hazard; small plastic bunting flags were used on the wire in public areas. or a high visibility tape could be strung along the top of the supports. There is also a risk of RF burns from accidental contact with a transmitting antenna; in the tests a good visual lookout was observed and transmissions were stopped if anyone looked like approaching the antenna. If you have a vehicle near the antenna it is well worth checking the manufacturer's specification for the maximum RF power to be used near it.

Although working with limited data, initial conclusions were that this antenna is successful for 5MHz contacts from 3 to 250 miles. The line of sight performance is best off the side of the antenna, whilst the NVIS is good at 10 to 80 miles. Lower angle lobes to the side of the antenna are good for ranges 100 to 200 miles from the elevated test site. Its theoretical efficiency must be in doubt due to its proximity to the ground; however the results speak for themselves – it works! On the elevated dry chalk, the nominal ground plane could be considered to be a long way down, making the antenna behave more like a free space antenna.

It has proved to be ideal for a quick set up and take down, possibly under emergency conditions and even in the dark. It is easy to transport and seems to have repeatable performance.

It is worth remembering that if the antenna is to be used in an urban environment some alternative to the ground spike will be needed, as these agricultural fence posts are not designed to penetrate concrete! It might be possible to scale the design up or down, for 40 or 80m, to test its performance on these bands. G3WYN has used his successfully on 40m by folding back each leg on itself to achieve the higher resonance. This antenna, on this band, can be recommended for emergency communications from almost any rural site over a typical county-wide distances.

Chapter 2
Verticals

KITE AND BALLOON-SUPPORTED VERTICALS

Vertical wire antennas are hardly novel, but even though Marconi used kites for his early experiments it is perhaps surprising that more amateurs don't use kites (or balloons) to support vertical wire antennas – particularly so on 80 and 160m where the length required is often too great for conventional ground-mounted supports. Provided the wind strength is in the range of 'light breeze' (4 – 6 knots) to 'fresh breeze' (17 – 21 knots), a kite is a good way of supporting a long wire antenna. The essential requirements for a kite used for this purpose is that it is simple, easy to launch, provides a good lift at low wind speeds and is rugged. When it is flown it should be stable; i.e. it should remain stationary in the sky even in a turbulent wind.

When using a kite to support an antenna it is advisable *not* to use the antenna wire itself as the kite support line: lightweight wire will undoubtedly break and heavy duty wire is likely to be too heavy to allow the kite to fly properly. Instead, use a lightweight stranded wire either wrapped loosely around the nylon kite line - see **Fig 2.1(a)**, or drop the wire vertically from below the kite, while still using the nylon line to tether the kite, as shown in **Fig 2.1(b)**. Note the use of a post at the ground end to

SAFETY FIRST!

Do *not* fly your kite near power lines or where it will fly over roads. Remember that the wind can change direction! Be *especially* vigilant about lightning – check the weather forecast before you go out and cancel kite trips if thunderstorms are forecast.

In the UK, the Civil Aviation Authority is the regulator for kites. No permission is required for heights of up to 60m above ground level, *unless* you are in aerodrome traffic zone (ATZ), where the maximum height is 30m. The CAA can advise if your proposed location is within an ATZ. If you wish to fly your kite higher, you must seek permission to do so a minimum of 28 days in advance – see www.caa.co.uk

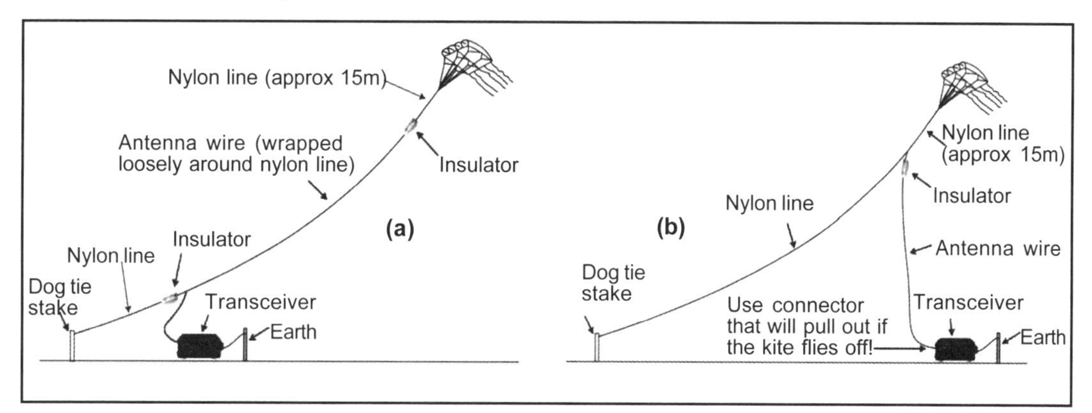

Fig 2.1: The two main methods of flying kite antennas.

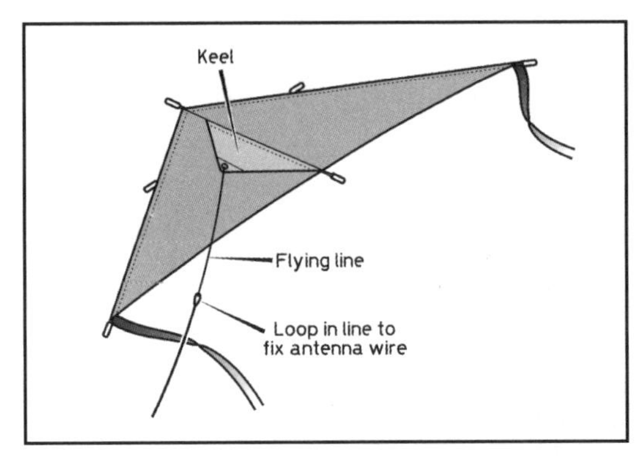

Fig 2.2: A kite suitable for supporting antennas.

give strain relief for the connection to the radio equipment: stakes sold for tethering dogs are suitable for this job. Include a connector such as a banana plug and socket in the antenna wire, close to the transceiver, so that it can easily be pulled apart. This will prevent the transceiver being dragged away should something go wrong. One disadvantage of the method shown in **Fig 2.1(b)** is that, should the wind drop, you end up with a lot of wire on the ground, as kite-flyer and noted portable operator Richard Newstead, G3CWI, commented in *RadCom*, November 2008.

Fig 2.2 is a close-up showing the sort of kite suitable for supporting antennas and the method of dropping a vertical wire from such a kite.

G3CWI reported that his favourite kite-supported antennas are a 5/8-λ wire for 80m or a 3/8-λ for 160m. However, note the height restrictions in the side panel 'Safety First!' All such vertical antennas will work best when fed against radial wires laid out on the ground.

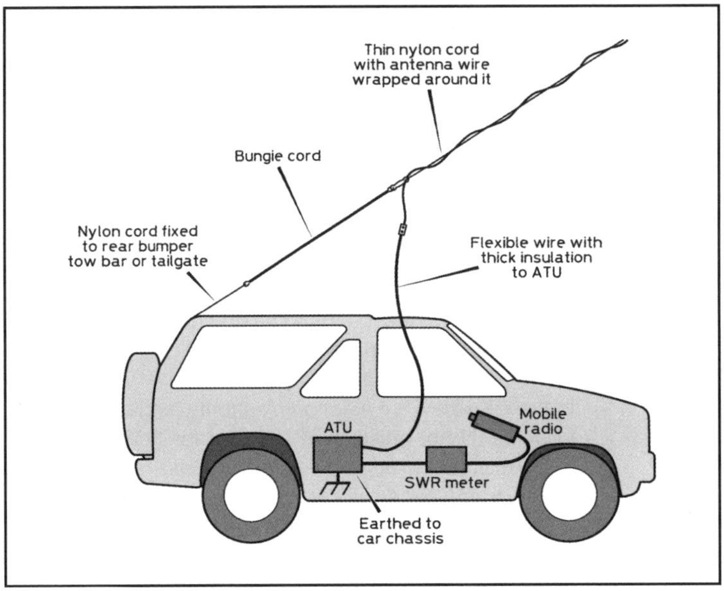

Fig 2.3: Method of connecting a kite or balloon antenna to a vehicle.

A vehicle provides a suitable 'shack' from which to operate when using kite or balloon-supported vertical wires, as described by Peter Dodd, G3LDO, in the November 1996 *RadCom*. **Fig 2.3** shows a typical set-up. The kite line is fixed to the vehicle as shown in the drawing. A length of bungee cord can be used to reduce the strain on the line and kite in strong gusty winds. The end of the antenna wire is brought through a wind-down rear window. When the wire is in position the window is wound up, making an effective insulated clamp. The wire must be positioned so that it is always slack, whatever the position of the kite; the nylon flying line should always take all the strain.

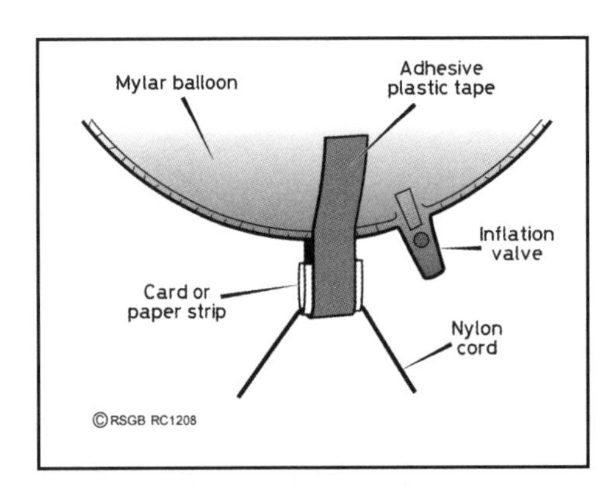

Fig 2.4: Method of connecting an antenna to a party balloon. The nylon cord is used for tethering several balloons together and the antenna is connected to the cord.

Unfortunately, the wind often disappears around dawn or just after sunset, when the DX is at its best on 80 and 160m – and just when you need a breeze to keep the kite aloft! A solution to this problem is to use a balloon – **Fig 2.4**. Mylar balloons are often obtainable from shops selling greetings cards, party shops, or from florists. The largest available is about one metre in diameter and is often referred to as a 'Jumbo' balloon. They have to be transported in an inflated condition (balloon gas is expensive). At least three or four of these balloons are required to lift a half-wave 80m wire antenna. An anchor point to the balloon is made using plastic tape as shown in the drawing. Balloons can be harnessed together via the plastic tape loops with a nylon cord and the antenna wire tied to the harness.

In zero wind conditions, when a balloon can be flown, the pull is very much less than that of a kite. The disadvantage of using several small balloons together is that the system has a lot of drag. Ideally, a balloon should be streamlined, like a barrage or advertising balloon, to increase the lift to drag ratio. The effect of drag is that even a very light breeze will blow the balloons sideways so that the antenna is anything but vertical. Nevertheless, a cluster of balloons is useful when the wind speed is 0 – 1 knot. Naturally, the antenna wire must be as light as possible, e.g. 0.6mm.

If the wind is too strong to fly a balloon yet insufficient to launch a kite, a combination kite / balloon is the answer. The 'Allsopp Helikite' is one solution and is particularly useful for supporting antennas. It is easier to launch than a conventional kite, with the helium providing the initial lift and the kite taking over once it is up in the air. They are available in various sizes – see www.allsopp.co.uk

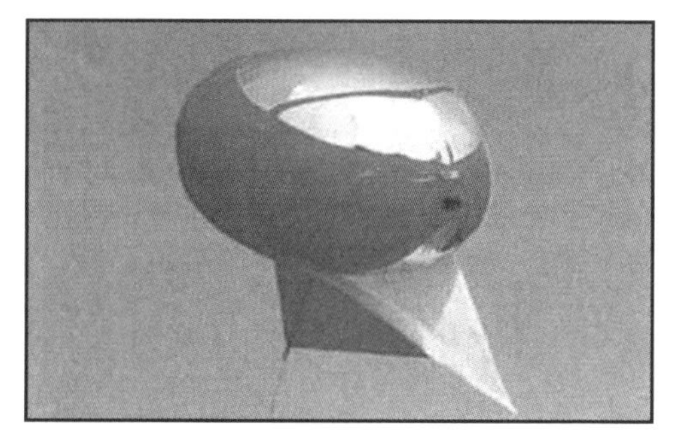

The Helikite, a combined lighter-than-air gas balloon and kite.

One disadvantage of using long vertical wires, whether supported by a kite or a balloon, is that of interference on receive caused by precipitation static. This occurs when rain, snow or even sand particles pick up a charge by passing through a naturally occurring electric field. This is discharged when they strike the antenna. Each discharge causes a 'popping' sound in the receiver and, with a large antenna, there can be many thousands of particles hitting the antenna each second. In bad cases, the intensity of the noise caused by the precipitation static will drown out all but the strongest signals. *Beware!* There may also be very large voltages present on the antenna that could damage the receiver or cause an electric shock to the operator. Connecting a 1MΩ resistor between the end of the antenna and earth will help to bleed the static discharge to ground: an earth system of one or more metal stakes in the ground and / or several radial wires will be required. However, it is safest to cease flying the kite or balloon altogether in such conditions.

HALF-WAVE VERTICAL BY G0KYA

Almost all 'homebrew' vertical antennas are a quarter-wave long, as this makes matching easy. However, it is also perfectly possible to make a half-wave vertical antenna and it is this that was described by Steve Nichols, G0KYA, in the October 2010 *RadCom*. As an end-fed half-wave has a very high impedance – around 3000–4000Ω – you cannot just connect the coax to the antenna and hope for the best: the secret is in the matching network. G0KYA acknowledges that the idea for this antenna came from the website of Steve Yates, AA5TB (www.aa5tb.com), which has a lot of information on the end-fed half-wave (EFHW) and other antennas.

The end-fed half-wave vertical is shown in **Fig 2.5**.

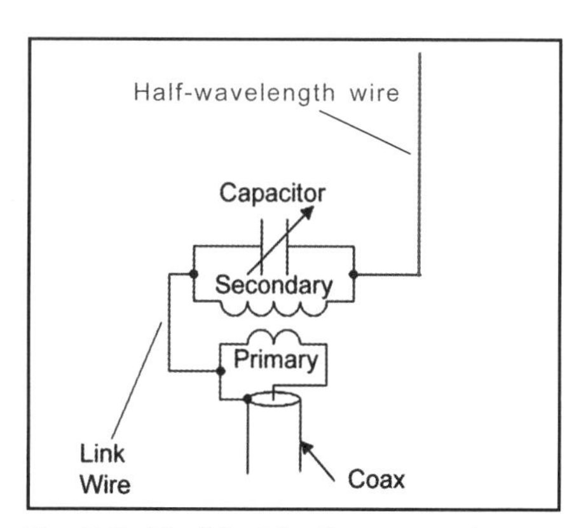

Fig 2.5: End-fed half wave antenna, showing the matching network. The capacitor is a short length of RG-58 coax.

The transformer is wound on a T200 (red) toroid. The coax capacitor is connected to the upper terminal block.

The half-wave of wire is taped to, or loosely wrapped around, a suitable fibreglass pole. The first EFHW built by G0KYA was for 10m. Take a half-wave length of wire, using the formula length (in feet) = 468/frequency in MHz. For 28.5MHz this is 16ft 5in (exactly 5.00m). Wind 17 turns of enamelled copper wire on a T200 (red) toroid as the secondary winding (each time the wire passes through the toroid is one turn). The T200 toroid is available in the UK from JAB Electrical Components (www.jabdog.com), and is also often available at rallies. The enamelled wire used by G0KYA was about 1.25mm diameter (18SWG) and is also stocked by JAB. Leave a little at the end for connections and then wind two turns over this for the primary, again leaving a little spare.

Connect a 24pF capacitor across the 17-turn winding using an electrical screw connector ('choc block'). A 10in (25cm) length of RG-58 coax can be used as the capacitor because RG-58 has a capacitance of about 28.8pF per foot.

Connect the feeder across the two turn primary and your antenna to one of the secondary wires. Finally, connect another piece of copper wire from the other secondary wire back to the braid of the coax (the link wire in the diagram).

The matching network should be built into a plastic box to waterproof it.

A quarter-wave vertical normally requires an earth or ground plane to work but in the EFHW vertical there is very little current flowing in the ground and it becomes almost unnecessary to have an earth connection. G0KYA found that this antenna can be fed without any earth stake, counterpoise or radials. The impedance is so high that little current actually flows down the braid. If you do get any RF problems, a coax choke of 8 – 10 loops of coax in a 6in (15cm) coil about a foot or two from the antenna should fix them.

To tune the antenna connect an antenna analyser to the end of the feeder and see where the antenna resonates. It will probably be lower than the 10m band. Snipping off half-inch (1 or 1.5cm) lengths of the coax capacitor will reduce the capacitance and move the resonant frequency higher. If the capacitor is cut down to about 4in (10cm) and is still resonant too low in frequency, remove one or two turns from the secondary winding. G0KYA ended up with 15 turns on the secondary and a piece of coax about 4in (10cm) long. It is better to remove turns than cut off too much from the coax. The final result was an SWR on 1.2:1 at resonance and less than 2:1 across the entire 10m band.

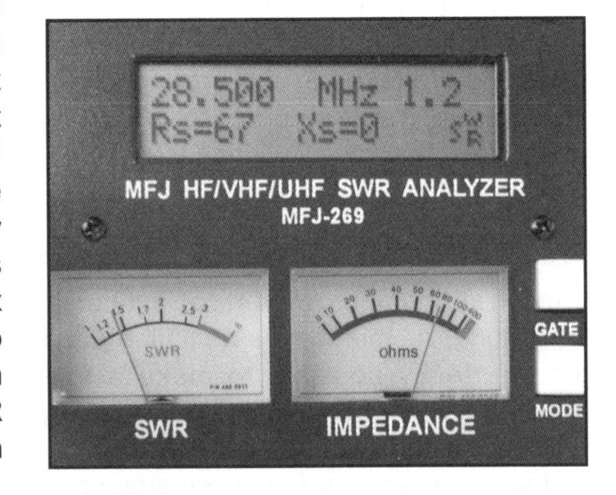

The EFHW vertical is easily scaled to other bands. By doubling the length of the wire radiator to 33ft (10.05m) you have an effective low-angle half-wave radiator for 20m. Using a calculator and the equation for the resonant frequency of an LC network:

$$f = \frac{1}{2\pi\sqrt{LC}}$$

it can be seen that halving the frequency means that the value of capacitance must be multiplied by four to make the antenna system resonant. G0KYA says, "I cut another piece of coax at four times the length of the original piece, hooked it all up and plugged it into the MFJ analyser. I couldn't believe it – the instant result was an SWR of 1.1:1 on 14.150MHz, rising to only 1.5:1 at the edges of the band."

The 20m EFHW proved to be at least as good as a half-wave 20m dipole at 30ft and US stations were "romping in" at G0KYA's location during the afternoon. The EFHW vertical outperformed a regular quarter-wave vertical with radials lying on the ground by a couple of S-points and was also a lot easier to put up. The antenna was used for Jamboree On The Air, when it outperformed a G5RV at 30ft by about one S-point. Later, the same 20m EFHW was used at an International Marconi Day special event station, where it functioned very well and stations around the world, including VK4, KP2 and numerous US and EU stations were worked.

THE WONDER WHIP

The 'Wonder Whip' is a home-made version of commercial products that go by names such as 'Miracle Whip', 'Wonder Wand' or Moonraker's 'Whizz Whip'. In the June 2006 *RadCom* John Goody, M1IOS, described how he built this antenna based on an original article by Robert Victor, VA2ERY ('The Miracle Whip – a Multiband QRP Antenna', *QST*, July 2001).

The Miracle Whip / Wonder Whip uses an autotransformer, rather than loading coils, to transform impedances in order to match a short telescopic 'whip' antenna over a wide range of frequencies for multi-band QRP portable HF operation.

The autotransformer may be considered as a double wound transformer; the bottom, primary part being connected to the rig, whilst the secondary is connected to the whip. The impedance transformation is the square of the ratio between the secondary and primary windings. The tapping point varies the ratio between the secondary and primary windings and establishes an impedance transformation.

The VA2ERY design used a continuously variable autotransformer, based on a coil transformer of 60 turns with the impedance ratio selected by a wiper sourced from an 'Ohmite' rheostat. Clearly there are considerable manufacturing difficulties in producing and winding the transformer core and the need for accurate construction and placement of a wiper mechanism. Using VA2ERY's autotransformer concept as a starting point, M1IOS experimented with a compromise design that would allow incremented tuning (rather than continuously variable), with the advantage of less-rigorous mechanical complexity.

The final design employs a 36mm OD ferrite ring wound with 50 turns of 16-gauge copper wire. The core is tapped at every second turn. Fine adjustment is achieved by altering the length of the whip.

From **Fig 2.6** it should be possible to assemble the various elements of the design, however the toroid requires

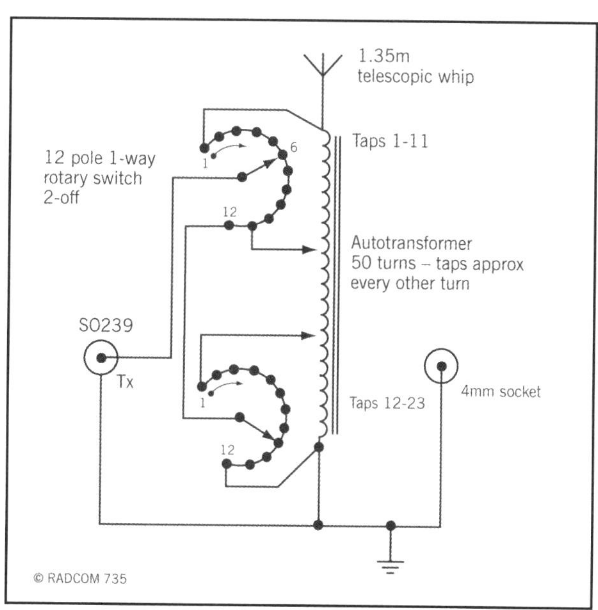

Fig 2.6: The Wonder Whip circuit diagram.

Inside the matching unit.

further explanation. The toroid core is wound with 48 turns of 16-gauge antenna wire. The core is tapped at every second turn. Because 24-way, single-pole rotary switches are not readily or economically available, two 12-way rotary switches are used, the 12th contact of the first rotary switch is connected to the pole of the second switch, to give 23 positions of adjustment. The 23 wire connectors to the toroid are directly soldered to the windings. This was achieved by:

- Cleaning and tinning the relevant core winding;
- Forming a small tinned loop to each connecting wire;
- Positioning each wire on the tinned core winding, and soldering the two tinned surfaces to fuse into a secure joint. This process requires patience and considerable care.

The telescopic whip is a Maplin LB10 HQ 10-section antenna 1.31m (4ft 3.5in) long, and 10mm diameter at the base. It has a 4mm hole to accommodate the antenna wire. A connector was made by drilling out the cable entry of a PL-259 connector to a 10mm interference fit into which the telescopic whip was screwed. The whip is secured to the PL-259 connector by finally soldering the antenna wire into position. A 3.5in floppy disc box is used as an enclosure.

The antenna is tuned by selecting the required band on the radio, and then rotating the two rotary switches on the ATU until the highest background noise or signal is achieved. The antenna peaks on receive and this will provide a good starting point for transmitting. Set your rig to a low power setting (5W) and transmit while observing the SWR meter. Rotate the rotary switch knobs until the SWR is minimised. Fine adjustments to the SWR may be made by altering the length of the telescopic whip (not whilst transmitting). A 10–12m (33–40ft) length of 1mm-diameter wire is used as a counterpoise, which simply lies on the ground.

VHF and UHF operation are performed by selecting the first position on the control knob and adjusting the antenna to a quarter wavelength. Horizontal polarisation, particularly for 144MHz SSB operation, is achieved by turning the antenna on its side.

On receive, performance is first-rate. S-meter readings appear very respectable. The best SWR figures that could be achieved using the 1.31m whip, both without and with a 12m counterpoise wire, are shown in **Table 2.1**. Does it *work*, though? Emphatically *yes*. On 40m M1IOS/P worked GB2IWM and received a 53 signal report while using 10W; SM4YPG gave a 47 to 57 report with just 5W; while on 21MHz AC5N in Oklahoma gave M1IOS/P a 59 report when he was running 15W to the Miracle Whip.

Freq band (MHz)	No counterpoise	With counterpoise
28MHz	1:1*	1:1*
24MHz	1:1*	1:1*
21MHz	2:1	1:1
18MHz	1:1	1:1
14MHz	1.5:1	1:1
10MHz	1.5:1	1:1
7MHz	4:1	2:1
3.5MHz	8:1	6:1

Table 2.1: Transmit SWR figures measured on transmit, using a 5W carrier (*after reducing the length of the whip).

I5TGC SMALL 80m VERTICAL

This very small vertical antenna for 80m was designed by Cesare Tagliabue, I5TGC. It was first described in *RadioRivista*, the official magazine of the Italian national amateur radio society, ARI, in October 1996 and later reported in the 'Antennas'

I5TGC's rooftop antenna 'farm'.

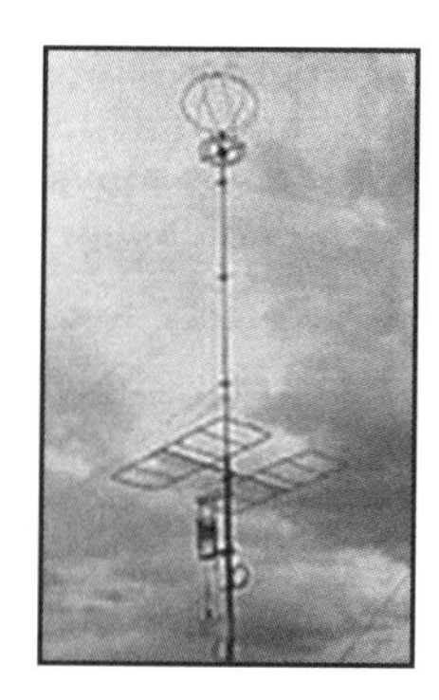

Fig 2.8: The 80m vertical, built by IK5PWN to the I5TGC design, with the 3.5m-long vertical radiator, a top capacity section, a top loading coil and two lower capacity frames.

Fig 2.7: The top loading coil showing its construction and method of connecting to the top capacity section and the vertical radiator. This is actually the 40m version but the construction is the same as for the 80m version.

column by Peter Dodd, G3LDO (*RadCom*, April 2003).

All I5TGC's antennas (including one for 136kHz!) are confined to the roof of his house in Florence (see photo above left). The most interesting aspect of these antennas is that apart from the 136kHz version they are vertical dipoles and so do not require any RF grounding. The 80m vertical dipole was originally designed and built in 1993. The example shown in **Fig 2.7** was built by Paolo Pieri, IK5PWN, and is made up of a 3.5m long vertical radiator, a top capacity spherical frame, a top loading coil and two lower capacity frames. A tuning and matching coil is used to tune the antenna and couple the antenna to the feeder.

The top capacity sphere comprises an open frame made from four 800mm diameter loops of aluminium tube. This is fixed to a short insulated section above the top loading coil. This coil has seven turns wound as a flat spiral with an inner diameter of 380mm and an outer one of 560mm as shown in **Fig 2.8**. The outside of this coil is connected to the top capacity section and the inside to the vertical radiator. The bottom end of the vertical radiator is connected to the bottom end of the tuning and matching coil, while the top end is connected to the two lower capacity frames. Each of these frames is 600 x 1500mm and constructed from aluminium tube. The tuning and matching coil is 170mm in diameter and has 20 turns of 4mm aluminium wire spaced at 8mm (see **Fig 2.9**). This provides

Total number of QSOs	1261
Number of stations	1122
Number of countries	120
Number of zones	27
Average signal reports	R4.95, S 9+2dB
Maximum signal reports	R5, S 9+40dB
Asia: AP, A4, A7, A9, DU, HL, JA, JY, OD, TA, UA9, UJ8, UM9, V8, XU, XY, ZC4, 4S, 4Z, 5B, 9K, 9V	
Average signal reports	R4.89, S6.87
Maximum signal reports	R5, S9+10dB\
Africa: C5, EL, J2, TU, 3V, 5N, 5T, 7X, 9G	
Average signal reports	R 4.9 S 8.26
Maximum signal reports	R5 S9+20dB
Americas: CO, FG, FM, HH, HI, HJ, KP4, LU, PY, PZ, TI, VE, VP2, VP5, VP9, W, XE, YV, ZP, 8R, 9Y	
Average signal reports	R4.84, S7.56
Maximum signal reports	R5, S9+20dB
Oceania: VK, ZL	
Average signal reports	R5, S7.25
Maximum signal reports	R5, S9

Table 2.2: From the digital log of IK5PWN, contacts made on the 80m band using the I5TGC 80m vertical during the period from October 1993 to January 1996. The signal reports indicate how DX stations received signals from IK5PWN and exclude Europe and '599' contest reports.

Fig 2.9: The tuning and matching coil with 20 turns tapped at 15 turns. The two-turn coupling link can be moved up or down the coil former rods for optimum coupling adjustment.

a greater inductance than required and coil taps are used for coarse tuning. The coil is coupled to the feeder with a two-turn link with a mechanical arrangement to allow for coupling adjustment. The band of 3.5MHz is fairly wide, so the antennas has a remote tuning facility. This consists of an aluminium disk, which can move along the axis of the coil by means of a threaded PVC rod, driven by a small motor and reduction gear. Both coils are supported using 15mm diameter MOPLEN rods drilled with 4.5mm holes.

So how well does it work? An analysis of QSOs made by IK5PWN during the years 1993 – 1996, using the I5TGC design, is shown in **Table 2.2**.

THE F4BKV VERTICAL DIPOLE ARRAY (VDA)

This is another antenna that could appear in more than one chapter – it is a vertical, but it is also a beam and it is based on a dipole. The antenna in question is the Vertical Dipole Array (VDA), which was originally developed in 2008 for the VP6DX Ducie Island DXpedition. Vincent Colombo, F4BKV, inspired by the strong signals received in Europe from VP6DX in the South Pacific, went on to develop a series of lightweight 2-element VDAs for his own Pacific island DXpeditions in 2012. More recently, Vincent's designs were further developed and used for the antennas on the very successful TX6G DXpedition to the Austral Islands in March 2014 (see the articles in *RadCom* June 2014 and December 2014).

The data here is taken from the original material on F4BKV's website at www.f4bkv.net It was first published, with F4BKV's permission, in an article ('Antennas for IOTA DXpeditions') in the *IOTA Directory* (50th Anniversary Edition, RSGB, 2014).

F4BKV's challenge was to make light antennas that could be transported easily by a single operator in his luggage. The antennas are based on 10m-long Spiderbeam (www.spiderbeam.com) fibreglass poles. The basic design of the antenna is shown in **Fig 2.10** and the theoretical dimensions for each of the five bands from 10 to 20m are given in **Table 2.3**. However, it is important to know the velocity factor of the wire used to make the antennas. For example, the CQ-532 stranded copper wire available from Spiderbeam has a velocity factor of about 0.96 which means that the wire lengths shown in **Table 2.3** must be multiplied by 0.96 to determine the actual physical length of the wires. The lengths are quite critical so do not cut the wires from the dimensions in the table without first applying at least an approximate velocity factor. (Further development by the TX6G team resulted in slightly different dimensions: these are given in the article 'Vertical Dipole Arrays (VDA)' by David Aslin, G3WGN, and Chris Duckling, G3SVL, in *RadCom*, December 2014.)

Spiderbeam CQ-532 stranded copper wire is used for the elements. The driven element uses a dipole centre insulator that allows direct connection of the PL-259 coaxial connector. No balun or current choke was used on F4BKV's antennas, though a ferrite current choke would be advisable, particularly if using high power. Several turns of plastic insulating tape are used to fix the dipole centre piece and coax to the boom to ensure they stay in place for the duration of the operation. The coax should be kept horizontal along the boom to the mast and then fixed vertically down the mast. Don't let the coax move by itself as it will affect the antenna's performance.

The reflector is made from a single length of Spiderbeam CQ-532 stranded copper wire, the middle of which is attached to the boom using insulating tape.

50 – 70cm of non-conductive string is attached to the ends of each of the wire elements in order to fix the elements to the mast while keeping them at their correct spacing (see **Fig 2.10** and **Table 2.3**). Do not tie them too tight: the mast should remain straight.

For the boom, F4BKV used parts of a cheap 6m telescopic fishing rod. The boom-to-mast mounting needs to be as light as possible yet strong enough to survive moderate winds. Details of the method used by F4BKV, as well as full constructional information, can be found on his website.

The antenna can be erected easily by one person if guy ropes are fixed in advance. One set of light guy strings is sufficient and they can be attached either at the boom-to-mast connection point or, preferably, above this. (Since the antenna has a broad forward lobe, you will probably not need to turn it.)

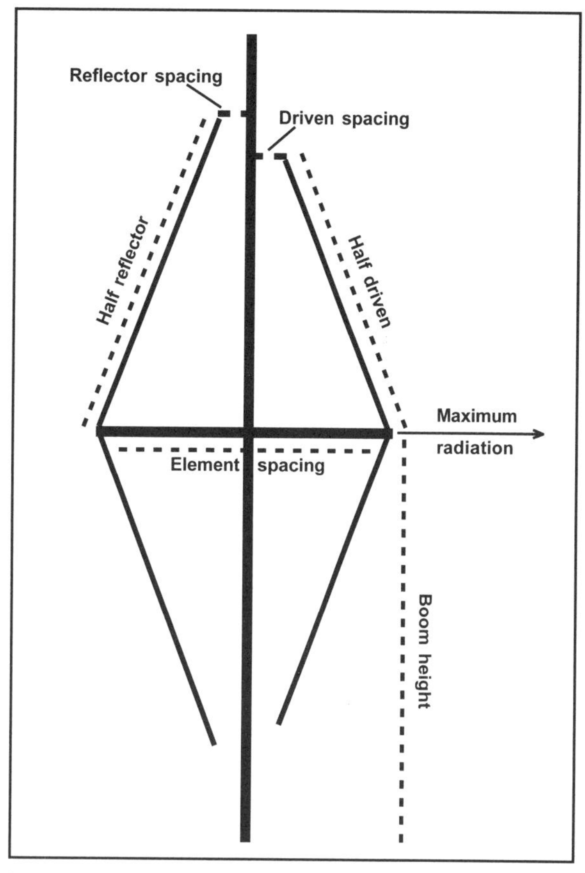

Fig 2.10: The basic layout of F4BKV's 2-element Vertical Dipole Array. For dimensions, see Table 2.3.

If the antenna is being installed on a sandy beach it is a good idea to dig a hole about 30cm deep and place the mast in the hole but, if you do this, remember to fix the boom correspondingly higher, or the antenna will be too low.

The lower ends of the driven element and reflector should be attached after erecting the mast (if you attach them while the mast is still lying on the ground, it will probably be too loose once in the vertical position). 50 – 70cm lengths of non-conductive string are used, the same as at the top of the antenna.

All vertical antennas need a good earth for the antenna to work efficiently. VDAs are no exception and, based on F4BKV's experience, he believes this design to be even more dependent on a good ground plane than other verticals. In other words, and to be absolutely clear: *if you do not plan to install the VDA near sea water, forget it and look for a different antenna design!*

Band (m.)	Element spacing	Boom height	Half driven	Half reflector	Driven spacing	Reflector spacing
20	3.56	6.24	4.97	5.23	0.46	0.42
17	2.78	4.87	3.84	4.09	0.37	0.33
15	2.38	4.45	3.31	3.50	0.31	0.28
12	2.02	3.91	2.81	2.97	0.26	0.24
10	1.78	3.66	2.49	2.63	0.23	0.19

Table 2.3: Theoretical dimensions for F4BKV VDAs (see text).

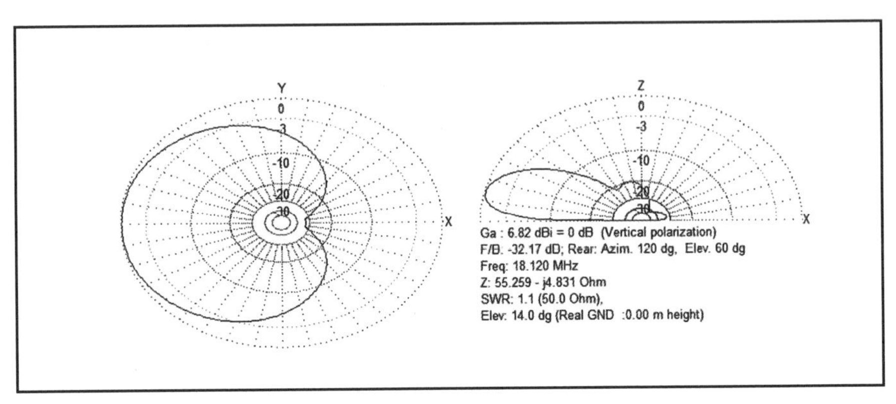

Fig 2.11: Radiation patterns of the 17m F4BKV VDA.

As can be seen in **Fig 2.11**, the forward lobe is broad, which is an advantage as you will not normally need to rotate the antenna. There is a sharp and deep null at 180° providing a good front-to-back ratio (about 20dB). F4BKV comments that he found this feature useful on islands where the antenna is beaming over the open ocean, with local noise coming from buildings directly behind the antenna being greatly attenuated. In a multi-transmitter environment, it also reduces interference from the other stations if the antennas are properly orientated. **Fig 2.11** also shows the relatively low elevation angle (14°) of the VDA.

Close-up of the F4BKV-design Vertical Dipole Array in use on Raivavae Island in the Australs group, French Polynesia (TX6G, March 2014).

Vincent, F4BKV, acknowledges the assistance of Cornelius Paul, DF4SA (the founder of Spiderbeam), as well as Jacques Saget, F6BEE, who carried out the antenna modelling using *MMANA-GAL* software (http://hamsoft.ca). It took a few days to assemble the first prototype but now each antenna can be put together and erected within about 15 minutes. Each one weighs just 1.5kg, including the 10m Spiderbeam pole, making the VDA ideal for IOTA DXpeditions close to the ocean, especially where weight is an important factor.

The article by G3WGN and G3SVL in the December 2014 *RadCom* provides a much longer and more detailed description of the VDA.

G8JNJ TC2M ANTENNA

The Terminated Coaxial Cage Monopole, or TC2M for short, is described as "a new design of broadband HF vertical antenna". It was designed and built by Martin Ehrenfried, G8JNJ, who has patented the design (GB2485812); the antenna may, however, be constructed by radio amateurs and used by them at their own stations with no restrictions.

This new design was explained in detail by G8JNJ in a two-part series in the May and June 2014 *RadCom*. What follows here are edited excerpts from the second part of the series, which mainly describes the practical construction of the antenna. Those interested in the theory and background to the design are referred to the original articles by G8JNJ, which can be found on the Internet at www.tc2m.info

The TC2M combines the best aspects of a 'fat' cage antenna and a Terminated Folded antenna, which make it possible to achieve a very wide instantaneous bandwidth and good efficiency without the need for a tuneable antenna matching unit. A simplified representation of the new design is shown in **Fig 2.12**.

The antenna consists of a five-wire 'cage' plus one centre wire. G8JNJ considers the five-wire cage to offer the best compromise in terms of overall radiation efficiency, ease of construction and cost of materials. Fewer than five wires results in much less consistent performance at the upper end of the frequency range, as the spacing between adjacent wires starts to become a significant proportion of a wavelength long, whereas using more than five wires provides very little additional improvement.

The central 'load' wire makes the antenna appear to be electrically longer than it actually is. This helps to improve the impedance match towards the lower end of the operational frequency range. The outer wire cage, in conjunction with the central load wire, forms a 'skeleton' coaxial transmission line. The impedance of the transmission line can be adjusted by varying the conductor diameter and spacing. This can be used to optimise the match between the radiating cage section of the antenna and a terminating load.

All six wires are connected together at the top of the antenna structure and the five outer wires are fed against a ground plane

Prototype Terminated Coaxial Cage Monopole, with wires emphasised for better visibility.

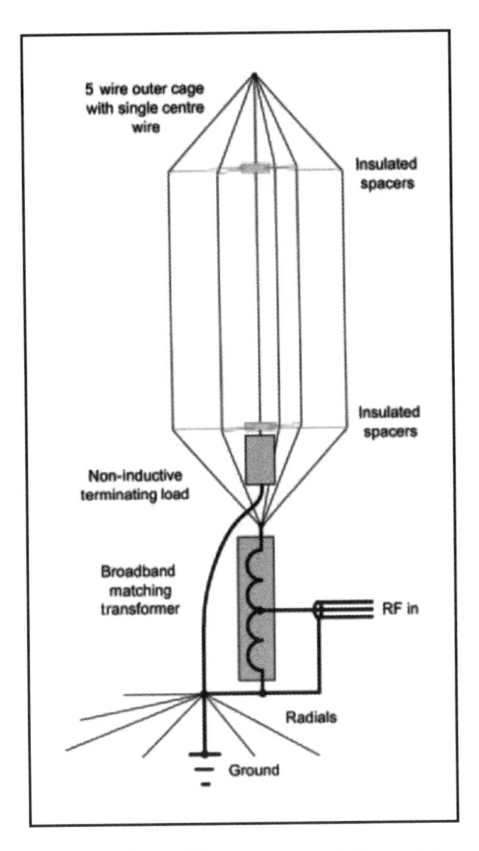

Fig 2.11: Simplified representation of the Terminated Coaxial Cage Monopole.

by means of a suitable unun at the base of the antenna. The central wire is connected to the ground plane at the base of the antenna structure, via a series connected terminating load.

Placing a terminating load at the end of the central loading wire, rather than connecting it directly across the secondary of the unbalanced to unbalanced transformer (unun), results in much less power being dissipated in the load and a better match throughout the operational frequency range of the antenna.

By making a suitable choice for the overall length of the cage it is possible to maximise efficiency over the required operating bandwidth. The upper frequency is limited by beam tilt when the antenna is greater than 5/8-wavelength long. The lower frequency limit is determined by the acceptable level of efficiency required by the user. A 10m long cage is capable of providing good performance over the frequency range 1.8 to 70MHz. A longer cage would be more efficient on the lower frequency bands, but performance on the higher frequency bands is likely to be degraded as a result.

The exact method of implementing the Terminated Coaxial Cage Monopole can be modified to accommodate different construction techniques or specific design requirements. G8JNJ has built versions using self-supporting telescopic GRP tube, guyed fishing poles and has also suspended wire cages from the limbs of trees. It may also be possible to use a rigid tube or tower to form the outer cylinder of the design, providing a suitable diameter centre wire can be found, to form a transmission line section of the correct characteristic impedance.

The low 'Q' broadband nature of the design means that it is not particularly susceptible to interaction with nearby objects. This makes it ideal for use in urban environments.

The five wires of the radiating cage are spaced by means of a central 'hub' with five spokes, all of which are made from a suitable dielectric, non-conductive insulating material such as plastic or GRP. When constructing the prototypes G8JNJ used plastic furniture castors to make the hubs. He held them in a wooden jig, which made it quick and easy to drill out using a standard pillar (press) drill. The photo shows a simple jig used as a drilling guide. The spokes were made by cutting up a cheap set of 5mm diameter GRP cable access rods and fitting sleeved grommets on the ends to help secure the wires and to prevent injuries to passers-by. By shopping around it is possible to build several sets of insulated spacers for under £10.

A simple jig used when making the insulators from castors.

The spacers are arranged to form a suitable support structure for the wire frame coaxial transmission line. By using 1mm diameter insulated wire with centre wire to outer wire spacing of 0.4 to 0.5m a wire cage coaxial transmission line, with a characteristic impedance of approximately 400 to 450Ω is formed. It is possible to use thicker diameter wire, but the wire to

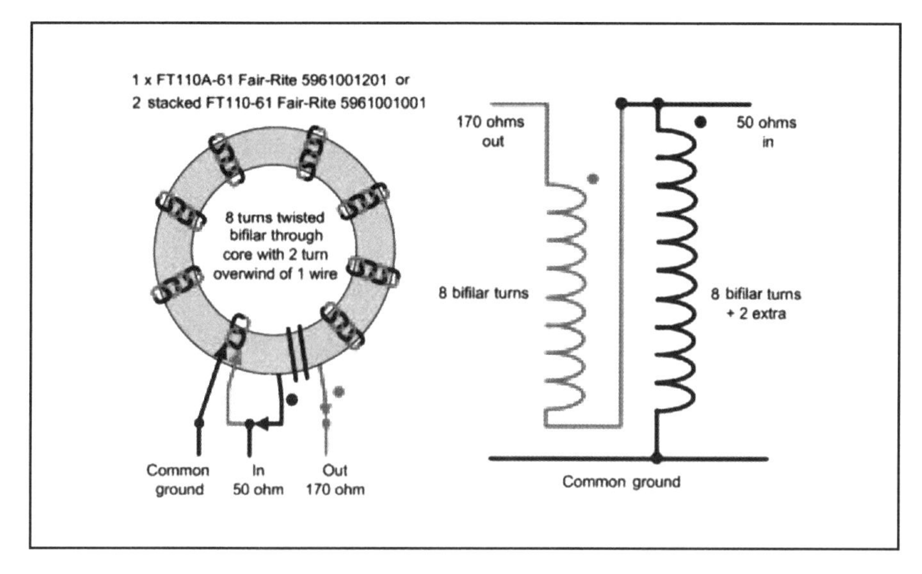

Fig 2.12: 170:50Ω unun construction.

wire spacing has to be increased in order to maintain something close to the target value of characteristic impedance. If you choose to use larger spacing between wires, you may also need to increase the number of wires forming the outer screen of the antenna because the effectiveness of the wire screen decreases as the spacing between the outer wires becomes greater than 1/10 of a wavelength at the highest operating frequency.

The input impedance at the feed point of the antenna is in the region of 150 to 170Ω. It is possible to use a standard design of 4:1 ratio Ruthroff (voltage) unun to achieve a reasonable match. However, in order to get the best results, it is preferable to use a non-standard ratio; although many constructors may consider that it is not worth the additional effort, it really doesn't take any more time to build. G8JNJ recommends the design shown in **Fig 2.12**, which was tested for extended periods with CW power levels of up to 250W. He used silver plated PTFE covered wire, but any reasonable diameter cable with good insulation would be acceptable. The choice of core size and material is critical. Do not substitute other types of ferrite or iron powder cores. If built correctly this design is easily repeatable, with a reasonably consistent impedance transformation and minimum amount of through loss.

One of the biggest challenges during this project was to source a high resistance, high power, non-inductive terminating load. Most non-inductive resistors are not suitable for this purpose as they only exhibit a non-inductive characteristic at frequencies below 1MHz. The power dissipation of the resistive load needs to be chosen to match the required transmitter power. For CW operation a wattage rating of 50% of the transmitter power should be used. If other forms of modulation such as SSB are used that have a lower duty factor, the wattage rating of the terminating load can be reduced accordingly. Also note that if there is inadequate heat sinking, or airflow, the overall power rating of any resistor may need to be reduced; especially if it is installed in a sealed enclosure, or mounted too close to other resistors in the bundle.

It is possible to modify the feed impedance versus frequency characteristics of the antenna by changing the value of load resistance. Computer modelling using *EZNEC* suggested that a resistance value of 450 to 470Ω, which is approximately three times the feed point impedance, would provide the best match across the required range of operating frequencies.

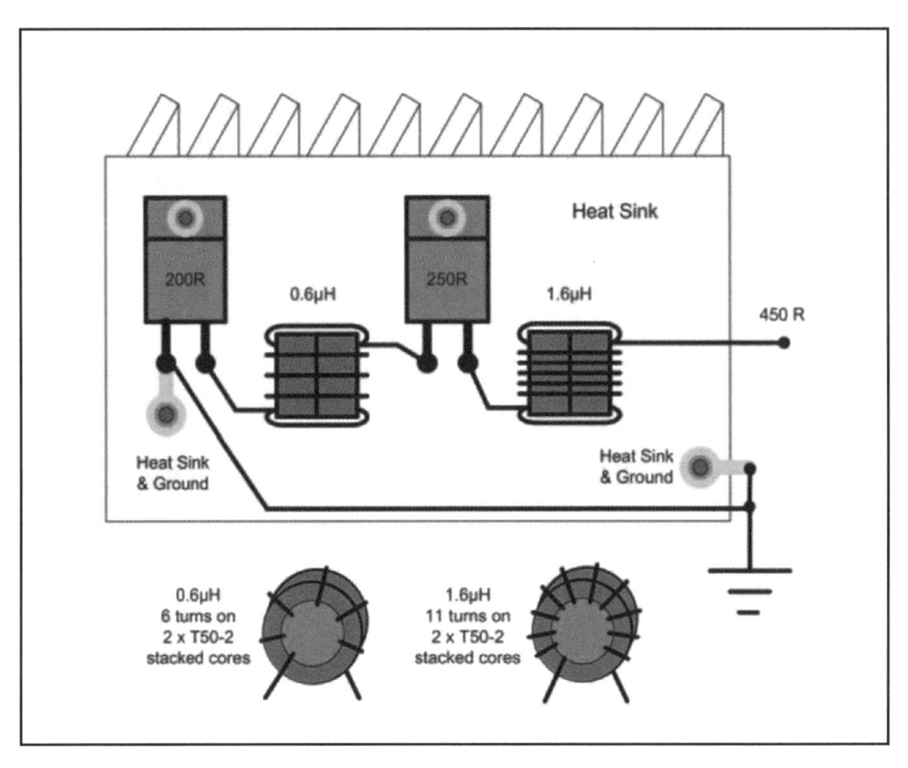

Fig 2.13: Detail of 450Ω terminating load and impedance correction network.

Because most resistors have a significant amount of capacitive reactance present when mounted on a heat sink, G8JNJ found it necessary to include two inductors in order to compensate in order to provide a satisfactory value of resistive impedance across the required frequency range. The configuration is shown in **Fig 2.13**. The full layout of the input transformer and terminating load, which is capable of being used with transmitter powers of up to 100W, is shown in the photo. Note that by building the whole unit in one box, which is also used to provide a heatsink for the load resistors, it is possible to configure the antenna quickly either as a conventional unun-fed cage monopole, or as a TC2M. This can be achieved by simply disconnecting the centre wire from the lower terminal and re-connecting it, along with the five outer cage wires, to the upper terminal. This feature is useful if you are concerned about the amount of power being dissipated in the terminating load, as it makes it very easy to compare the performance in the two different configurations.

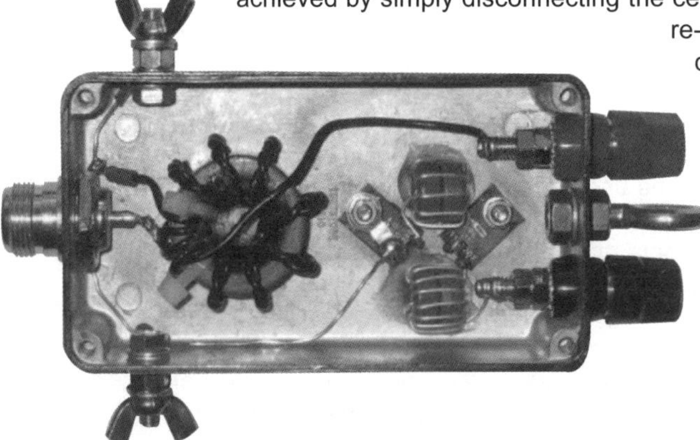

Practical realisation of the input transformer and terminating load.

As is the case with all vertical monopole antennas, in order to operate in an efficient manner this antenna needs to be fed against an appropriately dimensioned ground screen (ground plane, radials or counterpoise wires). A minimum of eight buried wires

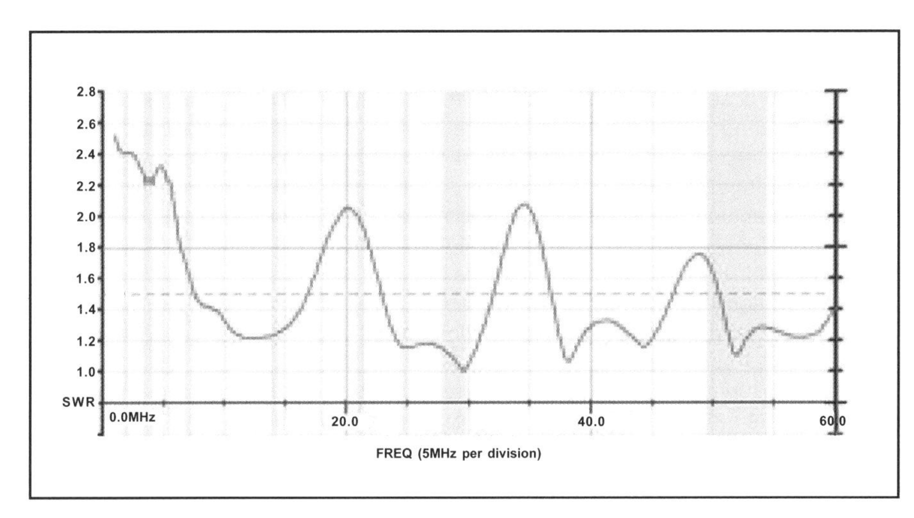

Fig 2.14: SWR from 1 to 60MHz, measured at the feedpoint. The horizontal line is at 1.8:1; the dashed horizontal line is at 1.5:1. The vertical grey columns represent the amateur bands.

would seem to offer the best compromise between cost, effort and effciency. If this is not possible then as many radial wires as possible should be used. If the wires are considerably shorter than 1/4 of a wavelength at the lowest required operating frequency, then it is better to use more wires. In practice, eight wires of 10m length with a further eight wires of 5m length laid in-between each other on the surface of the soil will produce reasonable results on most frequencies.

Although the antenna is designed for broadband operation, it may be that unwanted resonances are present in the radial wires. This is especially true if they are laid on the surface of the soil, in which case combinations of different lengths may be required in order to achieve a smooth impedance match across the required frequency range. Ideally, radial wires should be buried at a depth of at least 25mm in order to reduce the incidence of self-resonance.

Fig 2.14 shows the SWR over the operating frequency range of 1 to 60MHz. This has been measured directly at the antenna feed point, with no additional cable losses. Note that the SWR does not exceed 2.5:1 and in most cases is less than 2:1. This means that the antenna can be used without the need for a tuner over the entire frequency range. In a practical installation, a moderate length of coaxial cable will be required to connect the antenna to the transceiver. In such cases the SWR measured at the transceiver will appear to be even lower, due to the additional cable losses.

The main limit on effciency with the TC2M antenna is the amount of power dissipated in the terminating load and power wasted due to mismatch loss between the unun transformer and the antenna structure. Many engineers will naturally express concern about deliberately adding resistive loss into an antenna system. However, unwanted losses occur in all practical antennas. This can be through resistive or dielectric losses in cables, conductors, ground systems, matching networks and tuners. Although most designers would endeavour to reduce such losses, they can easily be in the region of 0.5 to 2dB, depending upon the impedance range presented to the tuner. In this design the unun is nearly always operating into an impedance that is close to its design value, so losses are greatly minimised.

On frequencies where the antenna is shorter than an electrical 1/4 wavelength, the resistive feed impedance of the radiating element decreases, and the mismatch losses associated with the unun transformer become much greater. Other losses also increase due to a greater proportion of the applied power being absorbed by the terminating load and ground resistance.

Calculated and measured gain differences show that at frequencies higher than 7MHz, where the monopole is greater than 1/4 of a wavelength long, the Terminated Coaxial Cage Monopole is almost as efficient as the cage antenna but with an auto-tuner. On some frequencies at the higher end of the operating range the TC2M is actually 1 or 2dB more efficient than using a similar length of thin wire and an auto-tuner. At frequencies below 1/4 wave electrical length the efficiency of the TC2M gradually tails off in a predictable manner, but it is still capable of providing useful operation at frequencies as low as 1.8MHz. In fact, tests on 160m have demonstrated a similar level of performance to that of a 100ft G5RV sized doublet (not connected as a Tee).

The measured performance at the lower end of the frequency range is actually better than the modelled values. This is not an error. The most likely explanation is that the system losses of the cage antenna, tuner and ground resistance are worse than calculated. So, by comparison, the TC2M results seem better than would perhaps be expected. This is not atypical of electrically short antennas on the lower frequency bands, as ground and tuner losses can be significant due to the low resistive and high capacitive value of the feed impedance encountered with such designs.

This design also functions very well as a wideband receive antenna. The reduction in gain at lower frequencies is not an issue, as the received signal-to-noise ratio tends to remain fairly constant, being dominated by external factors such as the location of the antenna relative to external noise sources, rather than its absolute gain.

G8JNJ concluded his articles by writing, “I hope that you have found this article informative and that it may have stimulated you to construct your own version of the TC2M antenna. I have found it to be very easy to build, as it is suited to the use of a variety of different construction techniques and materials. It can be made to be visually unobtrusive and not unduly influenced by nearby objects. It is therefore ideal for use in difficult or urban environments, where other designs may prove to be problematic. The simplicity of the design makes it easy to maintain. Whilst its performance is equal to, or better than, many commercial designs that cost considerable amounts of money.”

LIMITED SPACE VERTICALS

There is a tendency to think that a vertical is the ideal antenna for those with very small gardens. After all, the vertical radiating element takes up virtually no space at all, right? Unfortunately this thinking does not take into account the need for the radials to form the ground plane. Ideally, this should consist of a number of quarter-wave long wires extending like the spokes of a bicycle wheel in all directions from the vertical radiating element. Thus, for a 7MHz band (40m) quarter-wave vertical, the antenna would be at the centre of a square 20 metres long by 20 metres wide – not such a small antenna system after all.

Nevertheless, the vertical antenna remains an attractive antenna, particularly on the lower-frequency bands, as it provides the low-angle radiation required for long-distance DX contacts. A horizontal antenna would need to be mounted very high indeed above ground to achieve the same low angles of radiation as offered by a vertical.

Is it possible, then, to reduce the horizontal space requirement of radials without compromising – or at least without compromising too much – the performance of a vertical antenna system? The use of verticals with a short counterpoise (i.e. a single short radial) was investigated by the late Les Moxon, G6XN, in the 1970s and

1980s and a number of his ideas were published at the time in the 'Technical Topics' column compiled by Pat Hawker, G3VA, in *RadCom*. Writing in May 2006, G3VA commented that G6XN had emphasised that vertical antennas near ground are inherently asymmetrical and that it is almost always difficult to ensure that no common-mode current flows on the outer screen of the coaxial feeder. G3VA went on to say that he had been reminded of G6XN's work by a letter from Nick Brooks, G4BMH, who wrote that he had been working with the object of substantially reducing the length of elevated radials for an HF ground plane antenna, while keeping losses to a minimum.

G4BMH had devised a method of maintaining omnidirectional radiation with little loss in bandwidth or efficiency, but with a reduction in radial length to only 34%, or even 30% of that of two quarter-wave radials. The two conventional quarter-wave radials are replaced by one counterpoise conductor which is folded back on itself and extends back past the centre point: **Fig 2.15**. The diagram shows a 3in spacing between the wires but, in practice, 450Ω ribbon line can be used with only slight adjustment in the length of the open end. Note that the wire is a different length each side of the centre point – this is necessary to produce omnidirectional radiation.

The bandwidth of the G4BMH system is slightly less than with the reference two quarter-wave radials; however, it still comfortably covers the 7MHz band with low VSWR. There is some additional loss compared with the reference antenna (**Fig 2.16**): an additional 6Ω over poor ground; 5Ω over average ground and 4Ω over good ground. These figures are of little importance provided that the radiation resistance of the antenna is reasonably high (figures are for 7MHz, they will differ slightly on other bands).

All the 'rules' regarding elevated radials still apply: the height above ground of the radial system should ideally be about 4ft on 7MHz, 8ft on 3.5MHz, 15ft on 1.8MHz but G4BMH found that on 3.5MHz it works satisfactorily with the radial system on the top of a 6ft garden fence.

G4BMH wrote that an advantage of the system that may not be readily apparent is that as the currents in the earth system are in series, changes in earth conductivity or other conducting objects nearby will not alter the radiation pattern nearly as much as when two quarter-wave radials are used. It is well known that as the input impedance of a quarter-wave radial is extremely low, small unbalances in the radials can cause massive differences in radial currents, significantly affecting the radiation pattern.

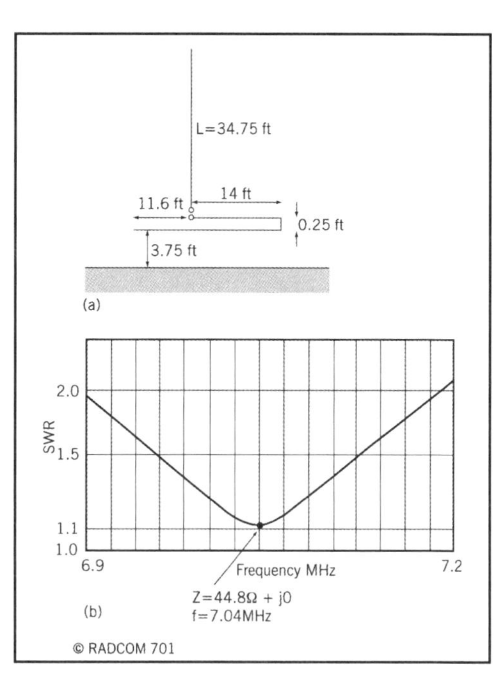

Fig 2.15: (a) 7MHz ground-plane antenna with G4BMH's elevated radial system suitable for limited spaces and not requiring the use of a ground stake. (b): Bandwidth simulation using *EZNEC*.

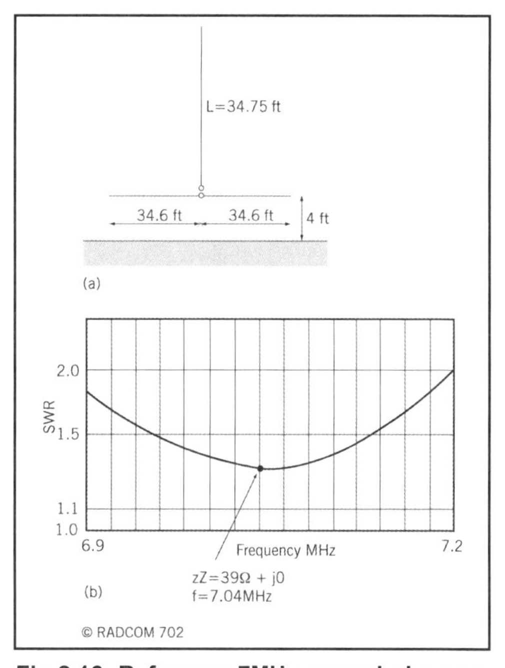

Fig 2.16: Reference 7MHz ground-place antenna with two quarter-wave radials and associated EZNEC simulation for somparison with Fig 2.15.

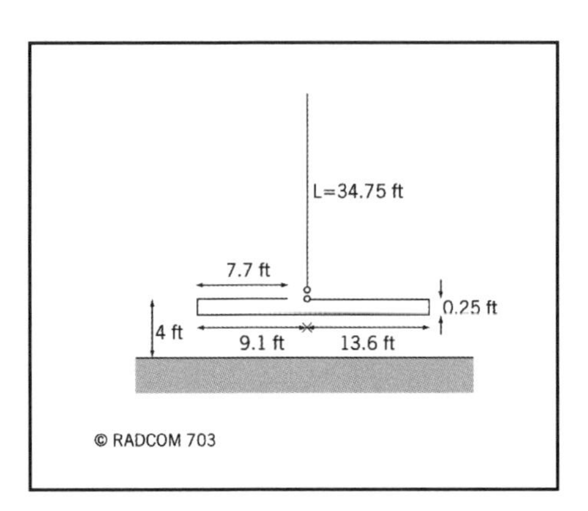

Fig 2.17: 7MHz ground-plane antenna using double folded elevated radial system.

The system can be scaled for other bands. The saving of space on 3.5 and 1.8MHz is enormous. There is only one 'hot' end on the radial, making for convenient adjustment. The length of the free end can be altered quite a lot before the radiation pattern, according to *EZNEC*, changes to any great degree, but the ratio of 1:1.2 seems ideal.

It would also seem possible to fold the free end back again, saving even more space (see **Fig 2.17**). In this case, the ratio of the free-to-folded length needs to be 1:1.5 and space saving is some 80%, but G4BMH wrote that his preference, unless space is really at a premium, is for the single-fold arrangement as this keeps the 'hot' end well away from the feed-point.

G4BMH also modelled an additional radial at right angles to the first. The bandwidth increased slightly and there was about 1Ω less loss, giving only a minimal overall improvement.

He concluded that an effective 3.5MHz DX antenna can be constructed on an average-size plot by mounting the earth system on top of a fence at the side of a house and, being of different lengths, makes them easier to accommodate in some gardens. The vertical element extends up to 40ft to a pole mounted on a chimney and along the apex of the house, and if necessary extending downwards again. Provided the horizontal portion is perpendicular to the radial system, it will radiate omni-directionally, with low-angle radiation.

Commenting on G4BMH's work, Dave Gordon-Smith, G3UUR, contributed the following to G3VA, which was published in the 'Technical Topics' column in the August 2006 *RadCom*.

G3UUR suggested raising the vertical above ground and then using two double-folded elevated radials (three parallel wires connected in series electrically as shown in **Fig 2.18**).

G3UUR went on to say: "Balance up the radials by trimming their lengths while monitoring the RF current in each using a clip-on current transformer.

"Fibreglass fishing poles are readily available these days in lengths up to 33ft [actually much greater lengths are also widely available – *Ed*] and would be a very convenient way of supporting a 14-gauge wire ground-plane antenna on 7MHz. Two 5m (16ft) fishing poles attached with conduit clamps to a wooden cross-piece at the base of the radiator could be used to support the folded radials if no convenient trees or posts were available.

"The spacing of the folded radial wires could be as little as one inch, or even slightly less. Since each radial is a single length of wire folded back on itself twice, the upper section could be taped to the fishing pole along its whole length. The middle section could be looped back from the far end of the pole, where it is very thin, and taped most of the way back along the thicker part of the pole. The bottom section could be slung below the other two using nylon cord at both ends to hold it in place.

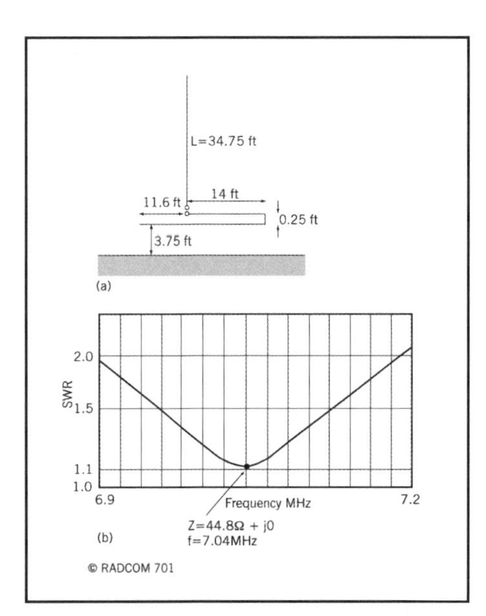

Fig 2.18: Quarter-wave vertical antenna with two linear-loaded radials for use where only a limited span (less than a quarter-wave) is available for the radials.

"In order to try to get the feed impedance nearer to 50Ω, the length of the radiator could be increased beyond a quarter-wavelength and the radials trimmed to restore system resonance."

COUPLED RESONATOR GROUND PLANE

The coupled resonator technique is not new but is mainly used in Yagi beams in order to provide resonances on two or more bands by using additional resonant 'driven' elements that are not connected electrically to the feedline. However, Joel Hallas, W1ZR, writing in *QST* ('Getting on the Air – A Folded Skeleton Sleeve Dipole for 40 and 20 Meters' *QST* May 2011, and 'The Folded Skeleton Sleeve Dipole on Other Bands' *QST* October 2011) showed how the same technique could be used to make a two-band dipole using window line.

He followed this up with another article, 'An Easy to Make Two Band HF Ground Plane' (*QST*, December 2013), again using the coupled resonator technique to provide, in effect, two antennas in the space of one – ideal for those with limited space who want to operate on more than one frequency band.

As with the original two band horizontal dipole, this antenna is really two parallel ground planes made from window line. The 20m and 15m bands were chosen for this model, however, it could easily be adapted to other pairs of bands. The lower frequency ground plane is fed directly with 50Ω coax, while no portion of the higher frequency ground plane is connected directly to the driven unit, but receives its energy through parasitic coupling (see **Fig 2.19**). While three, four, or even more radials could be used, the two radial configuration used is within 1dB of being omnidirectional and works very well for those with physical constraints.

Some have even used a single radial – picture a bent vertical dipole configuration – but that is more directional and has a higher angled main lobe, so the more conventional two radials are used here.

A vertical monopole can be ground mounted, but works somewhat better if raised off the ground. Another consideration is that the ends of the radials are a high voltage point, with a potential comparable to that at the top of the monopole. Thus for optimum safety, those ends should be high enough or otherwise protected from accidental contact. It is important to remember that while the radials serve as an artificial ground, in this configuration they are very much a part of the antenna and must be insulated

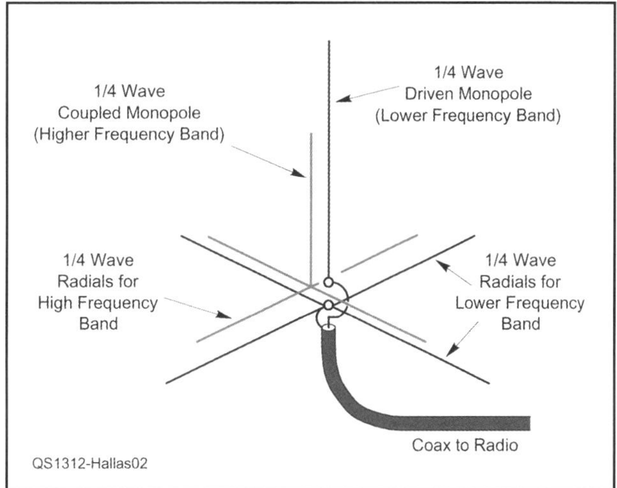

Fig 2.19: Simplified drawing of the two band coupled resonator ground plane. Note that there are no electrical connections to the higher frequency unit.

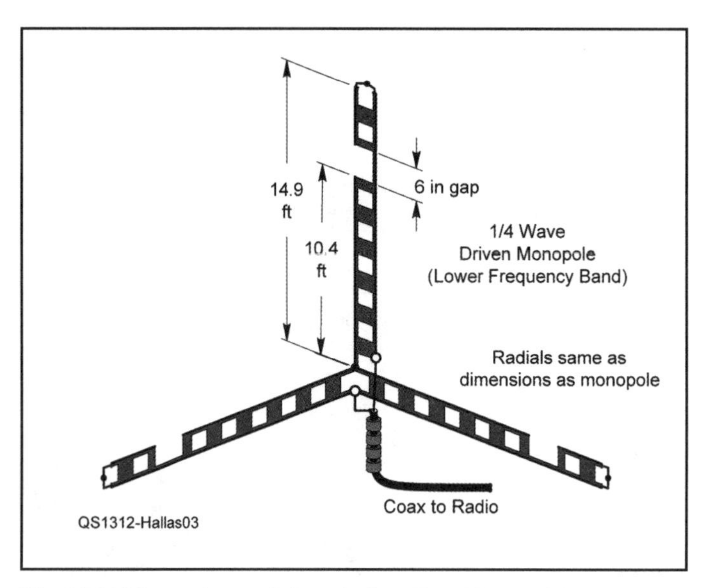

Fig 2.20: Dimensions of the window line 20m and 15m ground plane. For best match the radials should droop from the horizontal. While 45° is optimum, it will work satisfactorily with a less extreme droop.

as well as the vertical monopole.

If we were to have the ends of our radials at a safe height of 8ft off the ground, and droop our radials at 45°, we would need a support about 36ft above ground. If we had a single support that high, we would likely be better off with a dual band inverted-V for these bands. Thus, in most installations a lower height will be used and sufficient insulation and other protection should be employed, especially if higher power is used.

As with the coupled resonator dipole, it is possible to have either traditional linear elements or use the unused conductor of the window line beyond the end of the 15m elements to shorten the lower frequency antenna elements in a 'folded' arrangement. Since available height is a key issue and there really is no downside to this arrangement, a bit more than a foot can be saved by using the folded approach. Note that in the folded configuration, the highest voltage points are at the gap.

The exact dimensions will depend on the height above ground and the ground conditions. The dimensions given in **Fig 2.20** were set with the base 12ft above ground. The antenna was lowered so that the feed was at 3ft, and the radial ends were at about 1.5ft. As a result, the 20m resonance dropped by 270kHz, while the 15m resonance dropped by only 20kHz. Of course, this is a good direction in which to have a change, since it's easier to trim than stretch antenna wire. It is suggested to start with a few inches more than shown and trim as needed. The prototype was made from Davis RF window line that uses 18AWG stranded copper covered steel wire. While other types of window line could be used, different wire size or dielectric properties will have an effect on the final dimensions, so plan accordingly.

Note that while it looks like there are three distinct pieces, the two radials can be made from a single length of window line with the insulation stripped at the centre of

Close up of the feed arrangement for the prototype ground plane. The six ferrite beads that make the common mode choke are shown below the feed point. A permanent antenna would do better with more structural integrity at the feed. (The rope adjacent to the antenna is the halyard for a different antenna.)

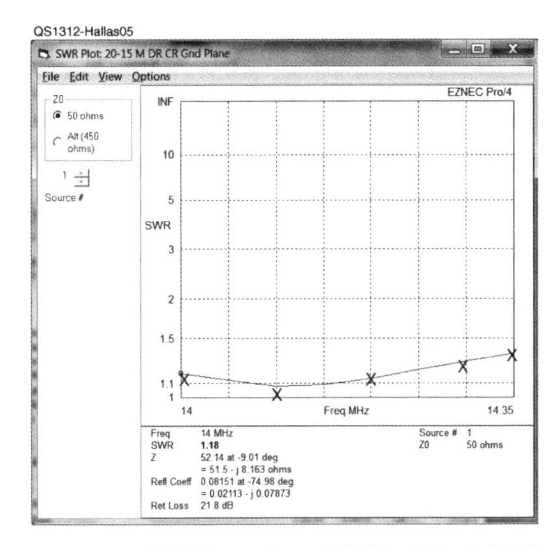

Fig 2.21: *EZNEC* predicted SWR of the folded coupled resonator ground plane on 20m. The measured results taken through 25ft of RG-8X are shown with the Xs.

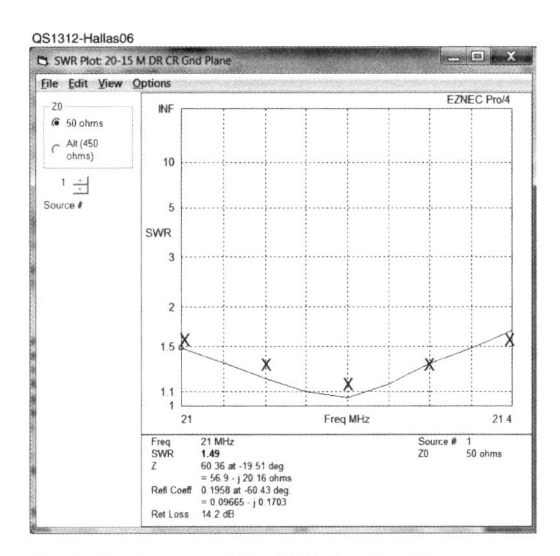

Fig 2.22: Same as Fig 2.21, but for 15m.

each piece. A common mode choke, made from six slip-on ferrite beads (Palomar Engineers FB-56-43, see palomar-engineers.com/ferrite-beads), is taped in place just below the feed. Of course, other kinds of chokes could be used, however, the beads noted slip nicely over RG-8X, don't require tight turns as on a toroid and are lightweight and compact.

While the design shows the monopole and radials the same length, if you are close to being on frequency, you can just trim the monopole. It makes the

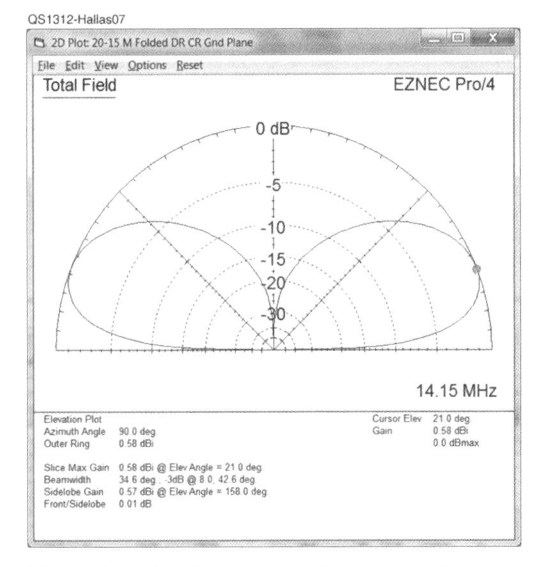

Fig 2.23: *EZNEC* predicted elevation pattern on 20m. The unfolded version (16.25ft high) at the same height has a gain only 0.1dB higher. The antenna was modelled with the base 12ft above typical ground. The azimuth pattern is within 1dB of being omnidirectional.

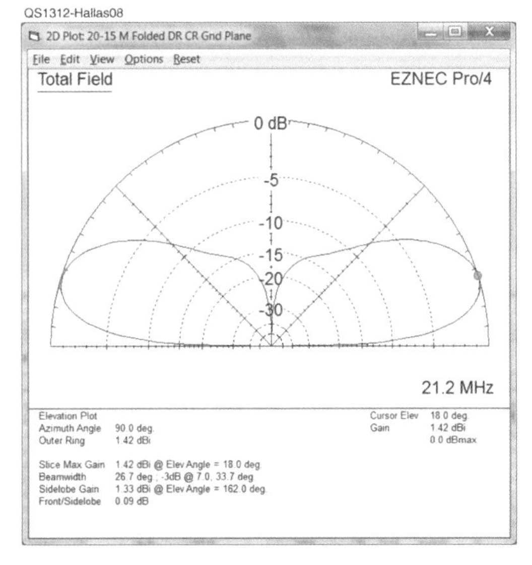

Fig 2.24: Same as Fig 2.23, but for 15m.

antenna the ground plane equivalent of an off-centre fed dipole. Any unbalance resulting from the small difference will be largely eliminated by the common mode choke.

The antenna came together quickly and works just as *EZNEC* predicted. The measured SWR for both bands at the end of 25ft of RG-8X is shown on top of the *EZNEC* SWR predictions in **Fig 2.21** and **Fig 2.22.**

Fig 2.23 and **Fig 2.24** show the elevation patterns on both bands, with the antenna feed 12ft above typical ground. Tests on the air tended to confirm the predictions. In A/B comparisons with a three-element triband Yagi at about 35ft, the Yagi always did better by around 3 S-units for stations at medium range that needed high elevation angles. DX stations were usually within 1 or 2 S-units of the Yagi. Part of the difference was due to the extra 2dB of line loss getting to the antenna test position. In summary, this antenna will not outperform a Yagi on a tower, but it will work DX at low cost and with a small footprint.

Chapter 3
Loops

FERRITE ROD TUNED MAGNETIC LOOP

Magnetic loops, or small transmitting loops (STLs), cannot be considered 'novel': they have been round a long time and provide a useful antenna solution for many who have little or no space for conventional wire antennas. So why am I including *any* magnetic loops in this book? The answer is that, in the standard design of STL, the metalwork and the tuning arrangements, in particular the variable capacitor, can be challenging for the home constructor – so novel approaches that address these difficulties are worth featuring. This chapter begins with two such innovative designs.

The first uses a ferrite rod to provide a variable inductor in order to tune the loop, in a similar manner to that of the 'VISTA' dipole by the same designer that is described in Chapter 1 of this book. John Seager, G0UCP, wrote about this loop antenna he designed in the September / October 2010 issue of the ARRL publication *QEX*. Like his VISTA antenna, this loop is designed for low-power (QRP) use only. It offers a new approach and is quite straightforward to make.

The schematic for a typical *conventional* magnetic loop is shown in **Fig 3.1(a)**. A loop less than 1/4-wavelength in circumference is fed with a gamma match. It is tuned at the high voltage end of the circuit with a capacitor that should ideally be of the vacuum variable type. A reduction drive is often fitted, as the Q is extremely high. You need an extension spindle if the capacitor has to be adjusted by hand, both for safety and to avoid affecting the resonance of the loop. All this can make for a cumbersome and physically fragile assembly, unsuitable for portable operation.

Now, if we could alter the resonance of the loop by varying its *inductance*, then instead of a variable capacitor we could use a preset or even a fixed type. The loop shown in **Fig 3.1(b)** does exactly that. It is tuned by moving a ferrite rod within a

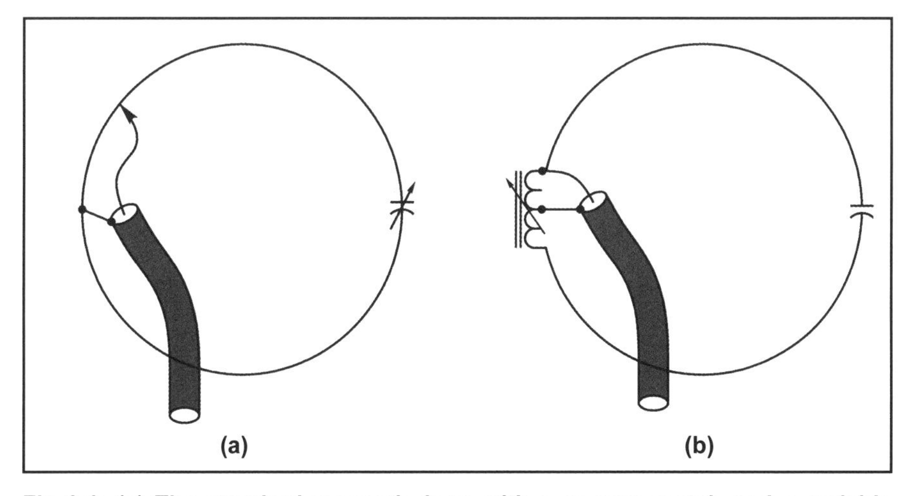

Fig 3.1: (a) The standard magnetic loop with a gamma match and a variable capacitor. (b) Magnetic loop with a variable inductance loop and fixed capacitor.

A variety of homemade capacitors to allow multi-band operation.

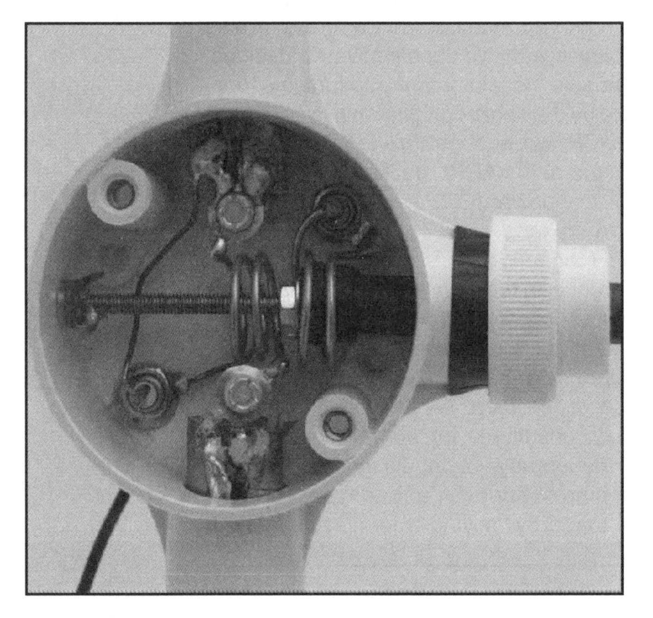

PVC T-box showing the adjustable ferrite rod and the coil in series with loop elements. The coil on the right is an inductive coupling.

small two-turn coil in continuity with the loop at the feed point.

The capacitor is a fixed one, chosen for the frequency range required, avoiding the need for any manual adjustment at the high voltage part of the circuit. The photograph shows homemade capacitors that permit multi-band operation from 3.5 up to 24MHz. This system also allows a simple way of coupling the loop to a 50Ω coax feed line. The traditional gamma match has a sliding contact from the coax inner core that is moved outwards from the zero potential until a match is obtained. In the present case this is unnecessary. A match can be achieved by connecting the coax braid to the centre of the two-turn coil (the zero-potential point), and the centre conductor to an end of the coil where it joins a main element of the loop.

This could be seen as an autotransformer or as a gamma match with part of the coil taking the place of the usual sector of loop circumference. If preferred, the loop can also be fed by way of an inductive coupling. It will be found that only two turns are needed for this matching transformer. Both coupling systems can be built in if each is brought out to a separate coax connector.

G0UCP made the elements out of thin copper tapes. These are very light but need support and so the loop was built from PVC conduit. Any other form of loop would be suitable, including the traditional solid copper tube, but the PVC T-box connector shown in the photograph is easy to work with and the whole structure is light and generally escapes injury even if dropped.

A ferrite rod of the type used in old AM portable receivers is fitted into a cup made from a rubber cable protector, or the plastic from a clip cover with a 2in (50mm) long 3.5mm diameter bolt threaded through and secured with a nut. If it is not firm, a drop of super glue helps. The end of the bolt passes through a 3.5mm jack socket that grips it quite well. The rod can be steadied at the open end of the T-box with a suitably modified cap from a hand wash bottle. A two turn 3/4in (2cm) diameter coil is connected between the two elements of the loop. The midpoint of this coil was the ground point for the gamma match. In the photograph a wire can just be seen connecting this point to the outer (braid) connector of the coax socket. As this is a low-power antenna G0UCP used phono sockets, but others may prefer to use a lower loss type. The connection to the centre conductor of the coax was soldered to a loop element where it is joined to the coil (either end will do).

The second coil is a simple coupling coil. Because the ferrite rod passes through this first (increasing the inductance as it goes) only two turns are required.

The elements for the magnetic loop were two double lengths of approximately 1.25in (3cm) wide copper 'slug tape' obtained from a garden centre. Due to the skin effect for radio-frequency current flow they should together be roughly equivalent to a tube 3/4in (1.9cm) in diameter. In one version G0UCP also added lengths of coaxial cable with the centre conductor and braid soldered together at the ends to increase further the effective diameter and reduce the conductive resistance. He made a square loop using conduit elbows at the corners and a (more-or-less) circular one in which the PVC was bent.

The square loop was made from 3/4in (2cm) conduit with 33in (84cm) sides joined with PVC inspection bends – curved sections of PVC with removable covers. The circular loop was made with 1in (2.5cm) conduit, with a diameter of 39.5in (1m). In both loops the fixed capacitor was mounted between two brass terminals in a PVC 'through box' (see photo) exactly opposite the T box housing the ferrite tuning assembly. All joints were secured with PVC cement on completion.

A capacitor (in this case for the CW end of 20m) mounted in the PVC 'through box'.

In this budget loop the only extravagance was in purchasing the copper tape and the use of silver solder, both justified by the need to minimise losses. Some of the capacitors can be custom made entirely at home from double-sided fibreglass copper board. The value needed for each band was found by substituting a large variable capacitor across the element terminals. The capacitance was noted at each required resonant point, using a low power signal source and a directional power meter. Capacitors were then fashioned as shown in the photo opposite. Copper board is readily cut to size if you measure the capacitance per area for the thickness you are using. If you go too far you can always solder a few pF back on the edge! On the lower bands a combination of capacitors and preset trimmers will be needed to achieve about 150pF for 7MHz and 680pF for 3.5MHz. Be aware, though, that voltages in small loops are high and inadequate capacitor insulation may arc over.

With an appropriate fixed capacitor in place, the inductive tuning arrangement gave full coverage of the CW end of each band. Separate capacitors were needed for the upper (SSB) sections on 14MHz and below.

Other than the fact that tuning is done with the ferrite rod instead of a variable capacitor, the procedure for setting up the antenna is the same as with a conventional loop. The feeder can be any length of 50Ω coax. To keep weight down in what can be a useful portable antenna 12ft (4m) of RG-174 was used. The antenna is kept within easy reach of the operator so it seems sensible to restrict power output to 5W or so. At this level no warming effect on the ferrite rod was noticed.

The coax lead is coupled via a bidirectional power meter to the transceiver and with power at a minimum the ferrite rod is rotated until a dip is seen on the meter. It will be found that if the jack socket spring contact is slightly crimped the bolt will track with almost micrometer precision and keep its position well. It is always possible to get a 1:1 match to a 50Ω output, if not with the preferred gamma match, then with the inductive coupler.

Subjective impressions are that this loop operates in much the same league as other magnetic loops. As with all loops it needs to be retuned when its height above ground or proximity to other objects is altered. Tuning is extremely smooth, with no hand capacitance effect even when touching the ferrite rod.

As G0UCP said, "The fun of making contacts with small antennas and low power is well worth the effort. Magnetic loops have been around long enough to find a firm place in the antenna handbooks. Is it time to try a new 'tune'?"

The completed square version of the G0UCP ferrite rod tuned magnetic loop.

STL WITH NOVEL TUNING CAPACITOR

The second small transmitting loop (STL) uses a different, but equally novel, approach to tuning the loop. Writing in his 'Antennas' column in the September 2010 *RadCom*, Peter Dodd, G3LDO, commented that the limiting factor in making an STL is the tuning capacitor. A good quality two-ganged or Butterfly transmitting capacitor or a fairly rugged vacuum variable is necessary, which has to be well engineered into the loop. Because the bandwidth of the antenna is so small, a method of varying the capacitor has to be built into the system. To try out an STL for himself, Peter built one using a capacitor arrangement made of hinged plates, as originally described by Martin Ehrenfried, G8JNJ, on his website at http://g8jnj.webs.com

The G3LDO loop is made from 22mm copper tubing in an octagonal configuration as shown in the photo above (the loop described by G8JNJ also used 22mm copper tubing, but in a square shape). Almost all of the material used to construct G3LDO's loop was obtained from a local DIY shop except the eight 45° couplings that were sourced from a plumber's outlet. A T-coupler was used at the base of the loop to provide a short stub mast for fixing the loop to an antenna support.

Most designs of STL use a 1m diameter loop to cover the bands from 20m to 10m, but in view of the state of the sunspots in 2010, G3LDO used a larger loop of

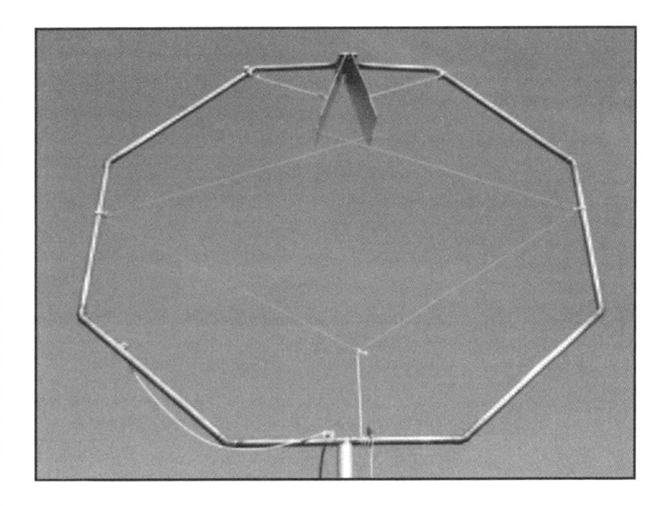

1.5m diameter which he hoped would cover from 7 to 22MHz.

The novel part of this design lies in the tuning capacitor, which is made from two aluminium plates fixed on hinges at the ends of the copper loop. Peter Dodd used 6 x 12in (15x30cm) plates because that is what he happened to have to hand. A drawstring and bungee cord arrangement is used to adjust the angle of the capacitor plates relative to each other, which in turn adjusts the value of the capacitance. The ends of the loop were flattened, which made a convenient point on to which to solder the brass hinges.

All descriptions of small transmitting loop construction emphasise the importance of overcoming the RF resistance of the capacitance to loop connection. This arrangement is no exception – the brass hinge probably presents a relatively high RF resistance, which is circumvented using coax cable braid as shown in the photo. Copper pads are used to make the connections to the aluminium capacitor plates. The capacitor plates held in the open position with 0.25in (6mm) bungee cord. Capacitor variation is achieved using strimmer line and nylon cord to pull the capacitor plates together against tension created by the bungee cord, which is best seen in the top photo. The bungee tension is found by trial and error. The strimmer cord is connected to the ends of the aluminium capacitance plates in a cross-diagonal manner using 22mm plastic tube clips. The strimmer cord runs through small holes drilled in these plastic clips.

Most loops use a motor / gear box arrangement to vary the capacitor and tune the loop. G3LDO used a simpler arrangement where the lower part of nylon cord section is wrapped around the lower part of the loop and secured with a plastic clip when the tuning point is found. This method of tuning was fine for testing the viability of the loop and was adopted because a suitable motor / gearbox was not available.

A shunt feed or Gamma match was used to feed the loop. G3LDO wrote, “I made a guess as to where to connect the shunt feed clip to the loop, connected the MFJ-259 analyser (set to 14.2MHz) to the feed point and pulled the cord of the tuning mechanism. The MFJ-259 registered an SWR dip of 1.5:1 on the first attempt. A small position adjustment of the shunt feed clip to the loop reduced the SWR to a much lower value.” Tuning was quite straightforward using an antenna analyser, alternatively you could tune for maximum noise or signals on receive and fine tune on low power using an SWR meter. The tuning arrangement performed reasonably well with just a bit of friction where the strimmer cord goes through the plastic pipe clip holes.

G3LDO calculated the maximum capacitance with the plates 4mm apart to be 100pF, which theoretically should tune the loop down to 10MHz. An insulator block is required to fix the distance between the two hinges. He used a block of dark coloured Perspex 10mm thick.

The completed capacitor is shown in the photo. Adjustment of the capacitance at the lowest frequency range proved to be rather critical with this tuning arrangement, because with the plate spacings changing from 4mm to 8mm there is a difference in capacitance of 50pF. The solution is to add a fixed capacitor in parallel with the variable one when using the antenna on the lower frequencies. This has the effect of 'bandspreading' the tuning. A short length of RG-213 coax was tried, as shown in the photo, and this worked quite well up to 100W (however, at 200W it flashed over). A better arrangement would be a fixed capacitance made from two aluminium plates fixed to the brass bolts and nuts. The minimum capacitance could be reduced by cutting off the two top (hinge) corners of each of the variable plates.

Before making the loop, G3LDO modelled it using *EZNEC 5*. The model was brought into resonance with a 24pF capacitance load, giving a 2:1 SWR bandwidth of 24kHz on 20m. The model predicted a maximum gain of 1.34dBi (including 22mm copper losses) compared with 2.2dBi for a loss-less free-space dipole. This model predicts that the loop will be around 80% efficient compared with a dipole.

The tuning range was not as great as the model suggested; the practical range covered only the 10, 14 and 18MHz bands. The reason is that the minimum capacitance of the hinged plate capacitor is greater than expected, although it was not possible to measure this capacitance with it connected to the loop.

This loop was tested on the 14 and 18MHz bands. Initial impressions were that it performed as well as a very good mobile antenna. The loop was mounted 2m above the ground, well away from the house, via a feeder comprising 43m of RG-213 and 10m of RG-58 coax. The comparison antenna was an 11m high multiband rotary dipole on top of the house, fed via 15m of RG-213. There was very little difference between the two antennas on short skip contacts. Sometimes the loop gave the best results, at other times the dipole did better. DX signals were a different matter, with the dipole 2 to 3 S-points ahead of the loop. On the other hand, the loop was often better on receive because it was so quiet.

SLINKY LOOP ANTENNA

The third magnetic loop in this chapter uses a conventional vacuum variable capacitor to tune the antenna but instead uses novel materials in its construction. Tom Haylock, M0ZSA, recalled reading an article by John Heys, G3BDQ, in the November 2003

Detail of how the Slinky is mounted on the water pipe. The string is tied to the Slinky on the hidden side of the pipe.

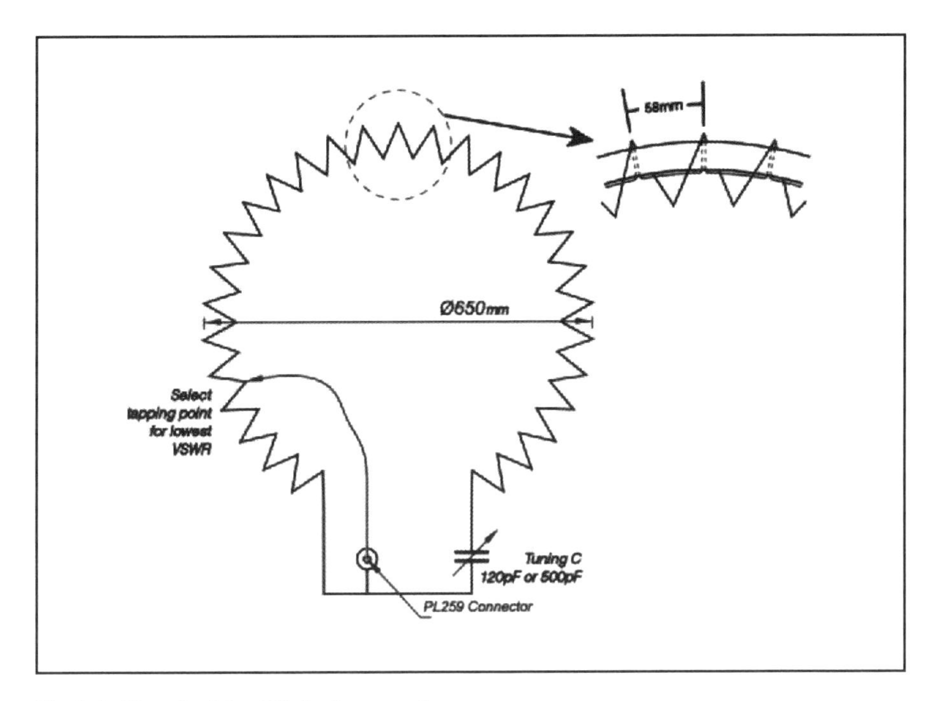

Fig 3.2: Circuit of the Slinky Loop antenna.

Practical Wireless about the 'Slinky Hula' antenna. G3BDQ's antenna used a 'Slinky' – the child's toy that consists of a metal coil that can 'walk' down a flight of stairs – and a hula hoop. Using G3BDQ's original design, M0ZSA built his own 'Slinky Loop' and described its construction in the November 2010 *RadCom*.

The circuit of the Slinky Loop is shown in **Fig 3.2**. M0ZSA's antenna is 650mm in diameter and covers 80 to 17m with an SWR of less than 1.2:1. It has been used at a power level of up to 100 watts. The antenna is made from 32 turns of a Slinky, 2042mm of 0.5in (1cm) plastic water pipe, a plywood base, strong string and, most importantly, one 500pF variable capacitor. M0ZSA happened to have a Jennings vacuum capacitor, considered to be the Rolls-Royce of variable capacitors, but if such is not available he suggests a variable with a slow-motion drive. He found that a 120pF variable did not tune down as low or as high as the 500pF capacitor and was very sensitive to the touch.

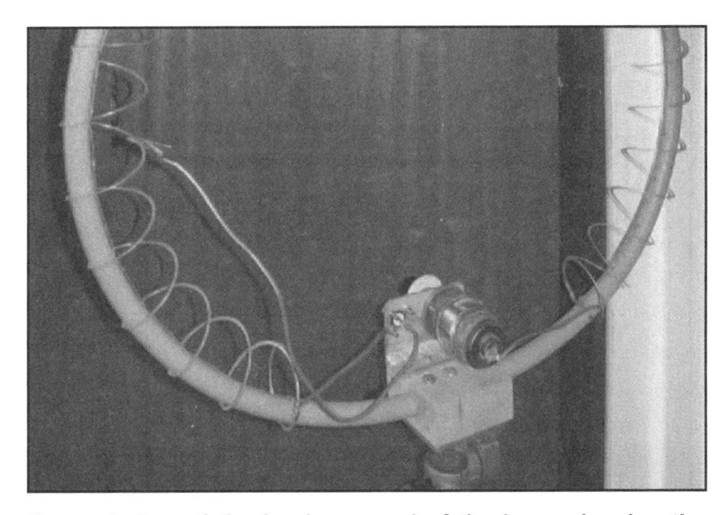

General view of the business end of the loop, showing the mounting block, capacitor / input socket bracket, wiring and tapping point.

The construction of the antenna is described in Tom's own words: "I started off with the pipe, marked the centre, 29mm either side of the centre and then at 58mm intervals 15 times either side of the centre. I then drilled 5mm holes right through the pipe (in a straight line). The Slinky is available from several sources. Mine came from Maplin in Brighton. Count 32 rings of the Slinky and cut it with a strong pair of wire snips. Screw one side of the pipe to a suitable or plastic base, making sure it is in the same plane as the rest of the holes. Thread the Slinky over the pipe, bend the pipe round and then screw the other side of the pipe to the base. Start with tying one full ring of the Slinky to the pipe and then one on each hole, ending with a full ring again (see photo).

"I made up a bracket to hold the input socket and vacuum capacitor (see photo). It doesn't need to be anything special, but there must be a good connection between the body of the input socket and the earthy side of the capacitor.

"It's important to find the right tapping point for best VSWR. To start with, connect the centre pin of the input socket about a quarter of the way round the loop. Tune the loop to resonance (listen on your receiver and adjust for maximum noise). Then transmit at low power, measuring the SWR. You will probably have to move the tapping point several times and maybe re-tune the capacitor because the settings can interact. However, once you've got a good match, you shouldn't have to change the tap for different bands. I used my MFJ analyser and found I could get a 1.2:1 SWR anywhere between 80m and 17m.

"Initial tests on 5MHz by Ken, G3WYN, got very good results. The loop was only a couple of S-points down on his full-sized dipole, which surprised us." M0ZSA says that a possible future development would be the ability to tune the loop remotely by using a stepper motor.

ALUMINIUM FOIL LOOP

One of the difficulties in making a successful loop antenna is the need to support a large circulating current. This comes about from the very small radiation resistance that this type of antenna exhibits. To realise any reasonable efficiency, the RF resistance of the loop needs to be in the region of milliohms. Many designs use a piece of copper tubing in an attempt to reduce resistive losses, ranging from one to several inches in diameter. The cost and weight of copper tubing is high – and is it really necessary to use such tubing and material?

The RF resistivity of any conductive material is affected by the changing magnetic field around the wire, which produces secondary currents known as eddys. The eddy currents set up in aluminium are not intense, partly as a result of the electron mobility behaviour of the atoms, which reflects in the permeability of the material itself. Aluminium has a DC resistivity nearly twice that of copper, but the resistivity to RF is almost the same.

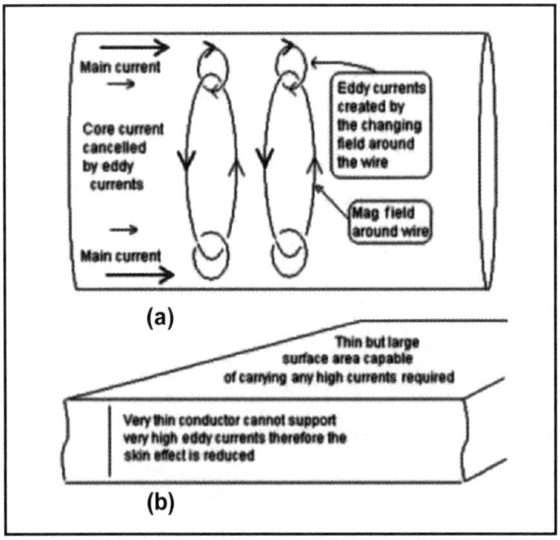

The object of any conductor carrying RF is that it must have the largest surface area that can be achieved from the smallest cross sectional area. This will enable a small RF resistance to be achieved. In an effort to clarify the use of aluminium foil as an efficient conductor for RF current, Andy Choraffa, G3PKW, decided to try very thin foil as a current carrying conductor.

Fig 3.3(a) shows the normal way that RF skin effect occurs in a circular conductor and **Fig 3.3(b)** the effect in a thin foil. In this latter case, the surface area of the conductor, that

Fig 3.3: Eddy currents a) in a normal wire conductor and b) in a thin foil sheet.

part that carries the RF current, is a maximum for a minimum of cross-sectional area. Such a cross sectional geometry can help to reduce the eddy current strength acting on the large surface area, in a similar way that the eddys in transformer laminations behave. The thin nature of the material geometry helps to reduce the transverse field effect of the resultant eddy currents from disturbing the main current, which is using the larger surface area to enable its transport. It is creating a geometry that denies the eddy current the means to have a lot of strength in that plane to disturb the main current.

A prototype loop antenna fabricated from aluminium foil. Note how its suspension points are well isolated from their surroundings.

G3PKW encouraged Roger Stafford, G4ROJ, to have a go at building a magnetic loop type radiator using aluminium foil. He fabricated a very neat looking system (see photo). Using such an antenna system, he demonstrated that a fat foil aluminium radiator is very efficient. It is capable of carrying high RF currents. Such small systems typically have very high circulating currents (15 to 25A) due to the very small radiation resistance. The loop shown was designed to be used on the 40m band. It was fabricated using a 10cm wide foil with an 8m circumference and a thickness comparable to standard cooking foil. Connections to the foil were achieved using a clamp arrangement fabricated from some scrap pieces of tube. Suitable holes were drilled in them and furnished with nuts and bolts, which secured a good ohmic connection. The ends were bent over and they in turn were connected to the tuning capacitor.

The loop was tuned to resonance using a 280pF butterfy type capacitor with a vane spacing of about a tenth of an inch, giving a resultant tuning value of 140pF rated at about 10kV in dry air. The two stators of the capacitor connected to each end of the loop. This tuned well on 7MHz and, when matched to the 50Ω coax line, achieved a unity SWR.

The coaxial feed was coupled into one side of the loop using a series matching 'gamma' capacitor. This capacitor is isolated from earth and has an insulating knob attached. The coax braid earth return was connected to the moving vanes of the butterfly capacitor, which became the system earth point. If the antenna was subject to any movement it changed its resonant frequency, which showed up on the matching bridge reading.

WARNING: when using such a loop antenna, please make sure to keep at least several metres away from the loop while it is transmitting. The circulating currents in the loop using 100W can be typically in the 10 – 20 amp region and the close-in RF field is very intense. *This must be taken seriously. Such a level field strength, although non ionising, may represent a health hazard.*

Operating the antenna at ground level gave comparable signal strengths to a full size dipole at 25ft. These were all with high angle signals. The signal from the loop was barely different from the dipole. 10dB of fading meant that the difference between the two was not easily measurable. The variation of signals was more than the difference.

Further tests with this system carried out on the Otterspool kite field next to the River Mersey gave similar excellent results. The antenna was supported just above ground using two short fibre masts. Wind blowing the elements made matching very difficult, though.

Building a similar system in a loft space would avoid the wind and maybe an even wider foil could be supported. Although not so easily obtainable in the UK, long lengths of 4in wide sheet aluminium are believed to be common in the USA, where it is used for roof flashing applications. This would no doubt be an ideal material to use, having enough strength to survive outside.

G3PKW commented that he has no doubt of the effectiveness of this type of system. The only difficulty is in the physical strength of such a system if used outdoors in the elements.

Thanks to Brian, G3GKG, and the 7157kHz group for providing many reliable and quantitative signal reports from around the British Isles. In fact all the members on 7157 were amazed at the loop's performance.

G3HBN PORTABLE MAGNETIC LOOP

Most magnetic loops are intended for fixed station use. However, for many years Jimmy Bolton, G3HBN, had used a portable magnetic loop for holidays and special event stations. The old design was simply a loop of RG-213 braiding slid over a piece of half-inch water hosepipe and supported by pieces of bamboo which formed a pear-shape. The whole was manually tuned and supported on a photographic tripod. It was time to upgrade the loop and G3HBN's new design was described by him in the June 2005 *RadCom*.

The requirement was to improve the performance and, if possible, increase the operating bandwidth. G3HBN wrote, "Looking at some aerial history, the old cage dipole came to mind. The cage dipole was designed to increase the bandwidth and help with the matching of it for commercial broadcasting purposes. In those stations, very long open-wire feeders from the transmitter to the aerial were customary. Aerial tuning units were not used and the feeder was coupled directly into the transmitter with either a link or a π coupling circuit. Such aerials would have a bandwidth of say 2.5 to 5.0 or 5.0 to 10.0MHz."

G3HBN's novel design for the magnetic loop was based along the lines of the old cage dipole. Instead of using a large diameter copper tube, as is normal with magnetic loop antennas, the element of this loop consists of 12 cables of stranded plastic-covered hook-up wire connected in parallel. The inner core of wire is about 1mm diameter. The overall diameter of the cage is 50mm. The element, when constructed, is placed on an hexagonal wooden frame. A tuning capacitor of 525 + 525pF with slow-motion drive is used to bring the loop to resonance at the desired frequency. The tuning range is from 6.9 to 32.0MHz. The loop is fed with a Faraday link coupling made with RG-213 coax, the braiding of which is open for 2.5cm at the centre.

One of the problems of magnetic loops is that they have a very narrow bandwidth. **Table 3.1** illustrates the comparison between a 1m, 22mm copper tube element and the 1m, 50mm caged element. A worthwhile bandwidth increase has been achieved. This represents an overall improvement in performance and is reflected in the results.

1m Portable loop 12-wire cage	Centre Frequency	1m Fixed loop 22mm copper tube
21	7015	11
30	10115	16
50	14050	30
75	18100	45
110	21100	60

Notes: All values in kHz. Measurements were taken at VSWR of 1.3:1 points with a Welz SP-300 VSWR / power meter. Both loops had a VSWR of 1:1 at the centre frequency.

Table 3.1: Bandwidths of the two designs of loop.

The outer diameter of the loop at the diagonals is about 105cm. When the length of the conducting element becomes greater than $\lambda/4$, the loop ceases to operate properly and becomes difficult to couple. It is desirable, therefore, to try to keep the overall conductor length to about 0.24λ or less at the highest frequency of operation. Although the model described here will operate on 29MHz, the element is really just a little too long.

3	old CDs glued together to form the centre core
2	plastic 3cm plumbing nuts glued together
7	plastic till-roll spools or similar rigid plastic tube that fits the dowels
6	dowelling spokes, 12mm, for the hexagon
5	plastic 10mm or 15mm wall-mounting water pipe clips
4	2.5cm (1in) rubber tap washers
5	3 x 40mm bolts with nuts and washers
5	water pipe saddles to strengthen centre mountings
12	55mm plastic discs
4	packets of 10m of 6A (24/0.2mm) hook-up wire (Maplin)
	Connectors to suit termination to the capacitor
1	525pF + 525pF variable capacitor
1	suitable plastic box or non-metallic container
1	slow motion drive 6 or 7:1 reduction.
1	well-insulated tuning knob, or plastic coupler and knob
1m	RG-213 Coax
5	Terry clips, 10mm
	Portable and rotary mounting

Table 3.2: Parts list.

Most constructors seldom follow exactly what is described in an article but, listed in **Table 3.2**, are the items that went into making this particular model. The assembly is fairly obvious from the photographs on the following page and from **Fig 3.4**. There are several points that are not so obvious, though. The overall diagonal measurements from the centre to the outer cables should not be less than about

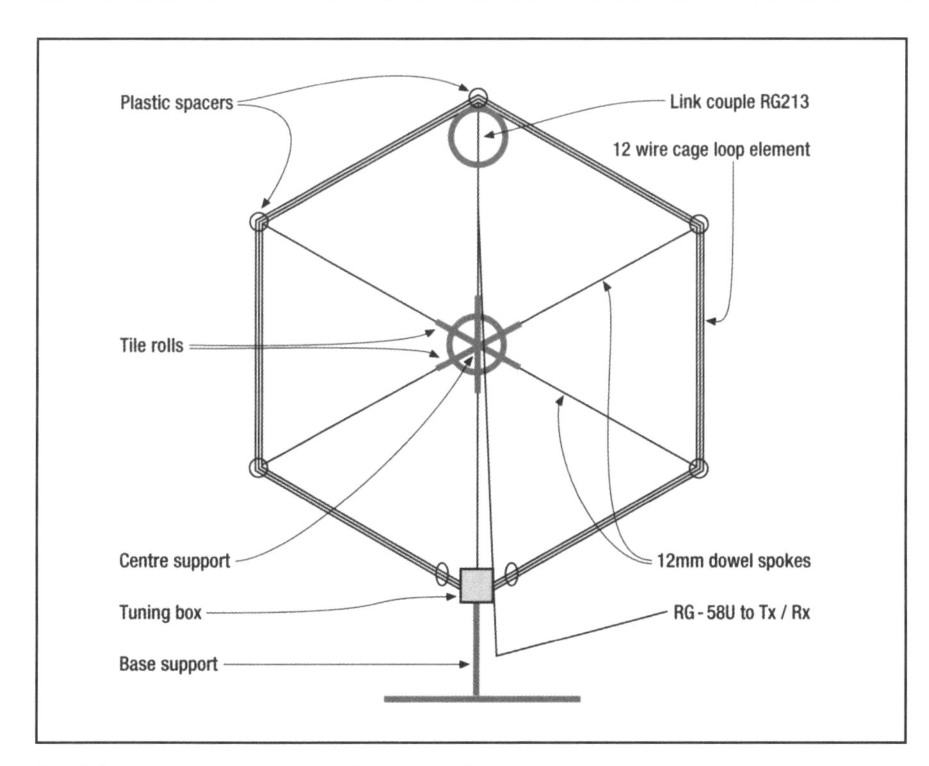

Fig 3.4: The magnetic loop – for dimensions, see text.

105cm. If the hexagon is smaller than this, with the cable specified, the loop might not quite tune to 7MHz. With this measurement, the loop should tune from 6960kHz to 32MHz. The dowelling should be cut into five lengths of 47cm and inserted into rigid plastic tubes (e.g. till-roll inners) at the centre hub. The sixth length is measured to fit whatever mounting box can be found. The seventh plastic support should be mounted to the tuning box and then the sixth spoke measured and cut. The five white plastic pipe clips are screwed to the ends of the five 47cm spokes.

The spacers are made with the 12 plastic discs cut from about 1mm thick plastic. A 3-litre food container was cut into flat pieces which were scored with 55mm circles. A further circle of 50mm diameter was scored inside each one. Marks should be made every 30° on the 50mm circle for drilling the holes for the cables. A centre hole should be drilled for the fixing bolts and rubber washers. The capacitor should be mounted and some suitable terminals used. PL-259 / SO-239 plugs and sockets were used in the example, but ordinary spade terminals would be easier and just as good. The capacitor is wired to the connectors and loop using only the fixed vanes, the rotor being left unconnected or 'floating'. This halves the capacity but doubles the working voltage. Further, it has the advantage of there being no moving contacts involved in the RF path. A well insulated knob, or plastic shaft coupler and knob should be used for tuning.

When the whole framework is assembled, the loop can be wired. Because the conductor is formed in a cage, the cables forming this cage will be of different lengths. Each wire must be threaded through the holes in the plastic spacers on the frame. The 12 wires can be cut roughly to length with plenty to spare for termination (coloured wires are a great help here) or from a reel of cable. But, in either case, each wire must be run separately, like stringing a musical instrument. The photograph below left shows the completed loop on a tripod mounting.

Start with the cables nearest the centre and work outwards to the top. Terminate all 12 cables at one end first and solder them to a lug or terminal or PL-259 etc. Connect this end to the tuning box. The other end is more difficult, because the wires need to be reasonably tensioned to form the cage. If they are a little slack, the rubber washers holding the spacers to the spokes provide some adjustment for this purpose. Once the loop element is finished, don't forget to mark which is the

The loop in use at G3HBN.

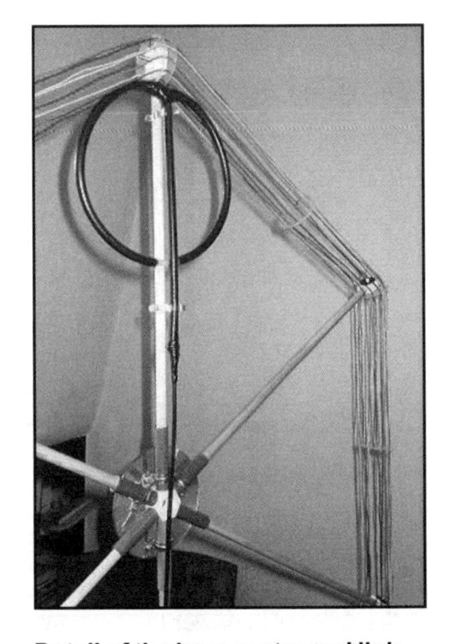

Detail of the loop centre and link.

top of the cage on the plastic spacers! Great care must be taken with all soldered connections to minimise any DC resistance.

The coupling link is made with RG-213. Many trials were conducted to optimise the coupling but, with the 'fatter' conductor for the loop element, it was found that the coupling link also needed to be 'fatter'. To form the link, strip about 4cm off the cover, expose the inner conductor for 2.5cm and solder the braiding to the inner. From the tip of the join, measure 60cm of cable. Strip about 3cm of cover to expose the braiding and solder the shorted end to the exposed braiding. This will form the coupling link of about 19 to 20cm diameter. At the centre of the link, cut the braiding for about 2.5cm to expose the inner core. Bend the remaining coax from the join to run vertically through the centre of the link and join it to the desired length of RG-58U. The mountings for the link are made with two Terry clips bolted back-to-back which clip neatly over the centre spoke and the RG-213.

The first step in tuning the loop is to peak it for maximum noise and / or signal strength on a receiver. If an antenna analyser is available, a quick check can be made for all the bands to observe the VSWR. A 1:1 VSWR should be obtainable on all bands from 7 to 29MHz. If 1:1 cannot be achieved, the antenna should be rotated, observing the VSWR at the same time. Adjacent objects in a room can unbalance the loop and, under these circumstances, it might not be possible to obtain the necessary VSWR. A VSWR / power meter is an asset, but the loop can be tuned with a simple field strength meter, tuning for maximum signal.

The tuning box.

Remember that maximum field strength radiation will be in the plane of the loop. On receive, this will be quite marked and a null will be obtained when the antenna is broadside to the signal. However, this does not always seem to be the case. Often very strong signals are received broadside to the loop and the same signal path seems to be effective for both transmission and reception. This may be due to building reflections, another illustration of where the loop needs to be easily and quickly rotated. This is particularly useful when operating with low power.

At the lower frequencies, the tuning of the loop is very sharp and it is essential to have a slow-motion drive fitted to the tuning capacitor. On the higher bands, the tuning is not quite so sharp, but it is still beneficial to tune the loop 'on the nose'. Hand-capacity has not proved to be a problem with this design. On the lower frequencies, the tuning capacitor is large and therefore any hand-capacity is insignificant. On the higher frequency bands, where hand-capacity is noticeable, it is still not too serious a problem because, at those frequencies, the usable bandwidth is much greater. There is, of course, absolutely no requirement for an ATU and if one is fitted to the equipment it should be either de-selected or tuned to a 50Ω load before using the loop.

This model is not built for high power, but it will comfortably handle 20 – 25W. Normally, any indoor antenna used in a built-up area should not really be used for high power, since the problems of RFI (Radio Frequency Interference) can become critical. The loop should be placed as far away from the operator as is practicable.

Tests were carried out using CW from London, and from a seaside cottage in Folkestone, with power levels of 5W and 20W. The majority of contacts made were at the 5W level. The overall feel of the antenna was quite amazing, with signals over

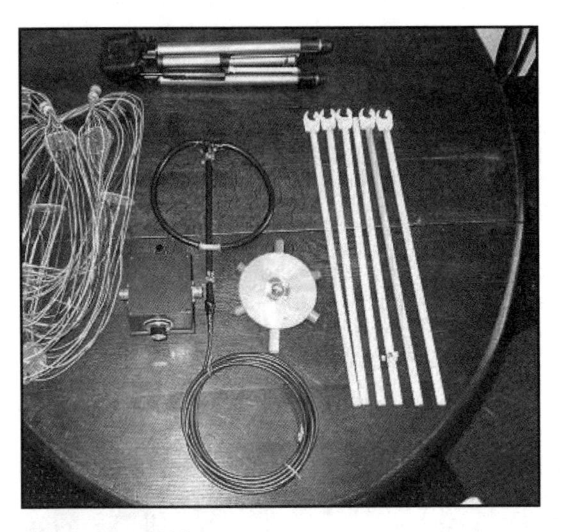

Ready for packing.

two S-units stronger than with the old 80cm loop, particularly on the lower-frequency bands. Comparison tests were made between the portable loop and the octagonal loop of 1m diameter on the roof. The roof antenna was, in the main, about 1 S-unit better, but the reports were that fading was more prevalent with the indoor loop.

Signal reports received varied with conditions, but reports of RST 579 and 589 were not unusual. Throughout the three month trial period, frequent contacts were made with most European countries, Asia and North America.

With 20W, the reply rate to stations called was between 70 and 80%; with 5W it was about 60 to 70% This is about the norm for QRP operation. Calling CQ was not very profitable (and seldom is with QRP). Many *two-way* QRP contacts were also made, one notably on 30m with GM3OXX (1W RST 579), G3HBN being in Folkestone (5W RST 599), the loop being at ground level in the sitting room, about 20m above sea level. This was a very long 'ragchew'.

Conditions throughout the test period have been at an all-time low and extremely difficult for making reliable evaluations. However, several DX stations were worked with QRP and that in itself was most gratifying. No tests were made for RF feedback when using a microphone.

The object of improving the original loop has been achieved with greater success than expected. It seems the application of a multi-cable radiating element for the loop has brought with it more benefits than originally anticipated. The increased bandwidth is far greater than expectations and the improved overall performance in the 'liveliness' of the antenna was a pleasant surprise. The next move is to replace the octagonal loop on the roof with a weather-proofed multi-conductor version. The results obtained here also open the door for further development in the general approach to the magnetic loop as an antenna in its own right, not necessarily to be compared with other antenna types. It is a radiator that has many characteristics that would seem not yet to have been fully exploited.

THE LAMBDA LOOP

After several HF small magnetic loops, we turn our attention to a horizontal full-wave loop – but not for HF as is usually the case for this sort of antenna. Instead, this is a design for a novel VHF mobile antenna, particularly suitable for use of 144MHz SSB. Geoff Grayer, G3NAQ, describes the development of the Lambda Loop as a practical mobile antenna that gives excellent performance. Writing in the March 2014 *RadCom*, G3NAQ says that his first experience of 2m mobile SSB operation, many years ago, had been using the original J-Beam circular 'Halo' antenna, later replaced by their 'Squalo' (square halo). These consisted of a gamma-matched half-wave dipole, bent round in a circle (or square). G3NAQ found that neither matched nor performed particularly well, and were not really mechanically suited to the fortitudes of mobile use. So he decided to look for an omnidirectional antenna with better overall performance. To this end he developed a set of criteria, listed in the panel below.

He first considered the 'clover-leaf' antenna, but decided that its large dimensions, mechanical complexity, and intrinsic fragility rendered it unsuitable.

THE IDEAL 2m SSB MOBILE ANTENNA

Electrical Requirements

Horizontal polarisation
Efficient radiator (implies a high-Q resonant antenna)
Good match to transmitter over band (144.15 – 144.4MHz in particular)
Omnidirectional horizontal coverage but limited vertical beamwidth, giving useful gain
Element(s) grounded at DC, reducing out-of-band signals and avoiding static build-up.

Mechanical Requirements

Low weight
Low wind resistance
Overall size appropriate, not overhanging car
Constructed from materials resistant to rust or corrosion
Strong enough to withstand normal driving plus wind speeds
Rapid assembly and disassembly for easy storage within vehicle boot
Robust, able to withstand frequent assembly / disassembly and storage.

Then he came across the Lambda Loop described in the 1983 *RSGB VHF / UHF Manual* [showing that a 'novel' antenna does not have to be a new design – *Ed*]. G3NAQ wrote, "Although I had never knowingly met anyone using one on 2m, nor ever seen one (it does not seem to be well known), I decided from its size and predicted performance that this was the one to try."

The finished prototype is pictured in the photo opposite and it satisfies all G3NAQ's 'ideal antenna' criteria. The Lambda Loop differs from the Halo by being one wavelength long rather than half a wavelength. Hence the name, after the Greek letter lambda (λ) used to represent a wavelength. It exhibits a good omnidirectional radiation pattern, which varies from equal to that of the Halo to about +7dB, >1 S-point (**Fig 3.5**). The exception is a deep null in the direction of the feed line, but according to this plot it is only about 10° wide at -3dB and 5° at -10dB, which would make it suitable for a direction finding (DF) antenna. However, when mobile one is always changing direction, so this is unlikely to produce anything more than very occasional additions to the variations in strength ('flutter') that are normally present on 2m mobile signals.

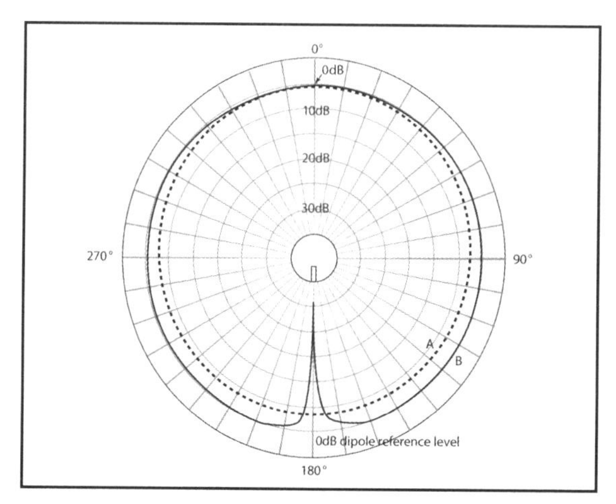

Fig 3.5: Radiation pattern of the Lambda Loop (solid line) compared with the Halo (dashed line).

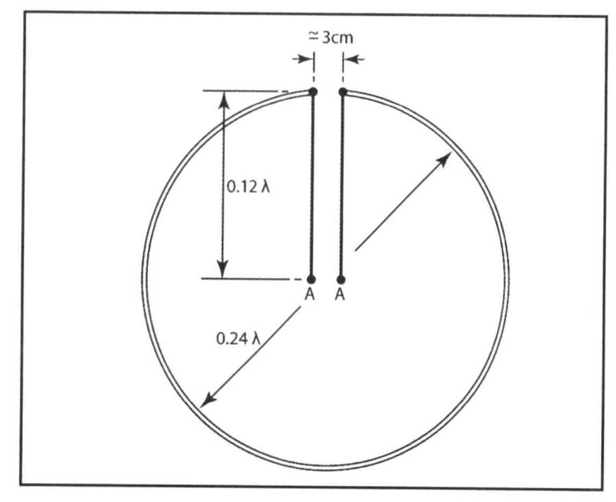

Fig 3.6: Dimensions of the Lambda Loop. Circumference = 0.7536λ, total loop size A to A ~λ.

The antenna length includes the parallel matching section (see **Fig 3.6**).The diameter of the Lambda Loop is 50% larger than the Halo, but it is supported at the centre, resulting in an antenna which is better balanced and (in my opinion) aesthetically more attractive.

The 1983 *VHF / UHF Handbook* states that the feed impedance of this design is 50Ω and recommends a 1:1 (λ/4 length) Pawsey stub as a balun (balanced to unbalanced feed) to connect to coax. This also serves as a convenient way of supporting the loop (it is sometimes known as the 'metal insulator'). However, no information on the dimensions or the materials used for construction was given. After building the loop as pictured using the materials and dimensions described, the impedance was measured at the end of a length of coax one (electrical) wavelength long, to ensure that the value measured was the same as that at the feed point. It was found to be much higher than the claimed 50Ω. Fortunately, the Pawsey stub can be used as a transformer to match almost any complex impedance by adjusting the position of the tap point and its length.

Since the horizontally polarised antenna will primarily be used in the SSB section of 2m (144.1 to 144.4MHz), the SSB centre of activity frequency of 144.3MHz was used as the design frequency. This translates into a wavelength λ of 2080mm (dimensions are rounded off to the nearest mm or so, as they are not so critical). The diameter of the loop is specified (**Fig 3.5**) as 0.24λ = 500mm, thus the length of the two radial sections are 250mm. The circumference is specified as 0.7536λ = 1566mm. The overall length of the loop plus two radial lengths is 2066mm. Because a 2m length of aluminium was used to construct the loop, there is 66mm missing. Hence the radial sections are made 217mm long, and the extra 33mm is provided by the supporting brackets (see **Fig 3.6**).

The length of the Pawsey stub was found empirically to be 215mm, but the total length is specified as 250mm, to allow for some adjustment. The coax is connected 138mm from the top. For protection and convenience, the coax is fed up the centre of the support tube, an angled hole being drilled to bring out the coax about 10mm below the tapping point (to allow for some adjustment).

No high level of skill is needed for its construction. You will need a number of hand tools (drill, hacksaw, files, hammer, pliers, ruler and scribing tool, and abrasive paper) and a vice. Round off all corners and sharp edges with a file and emery paper; you do not want any possibility of accidental injury or damage when handling the antenna. **Table 3.3** shows the materials required.

In the mobile environment, where the antenna is probably going to be regularly mounted and dismounted, something that is not easily damaged and would spring back if bent, i.e. a pre-stressed loop, is required. Strip rather than a tube is more appropriate for this design. G3NAQ came across 2m x 25mm extruded hard aluminium strips, intended for decorative purposes, at B&Q. These are ridged on

1 pc	**Aluminium strip 2m x 25mm trim (B&Q).**
1 pc	**Aluminium tube, hard drawn, thick walled, length optional (see text), x 12.5mm (or near) diameter**
1 pc	**Aluminium sheet (hard), 120 x 80 x 1.5mm, cut into 4 strips 20mm wide.**
1 pc	**PTFE block (or see text) 60 x 30 x 10mm.**
Use rust-free material for the following:	
2 pc	**M5 x 10mm bolts, plus wing nuts and lock washers (2BA could be used).**
8 pc	**M3 x 5mm bolts, plus wing nuts and lock washers (4BA could be used).**
6 pc	**M2.5 x 10mm bolts, plus wing nuts and lock washers (6BA could be used).**

Table 3.3: Materials required.

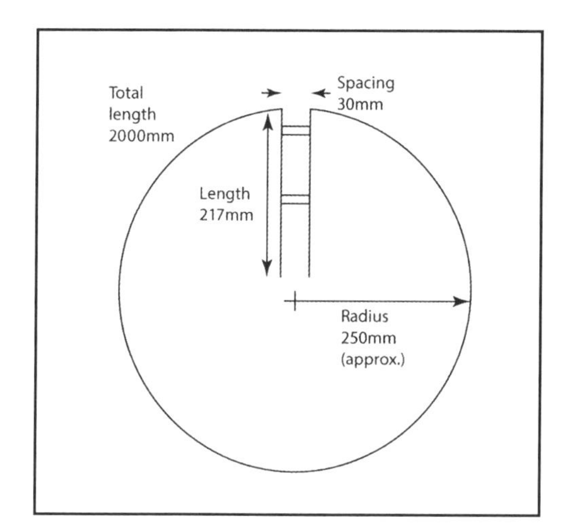

Fig 3.7: Loop dimensions.

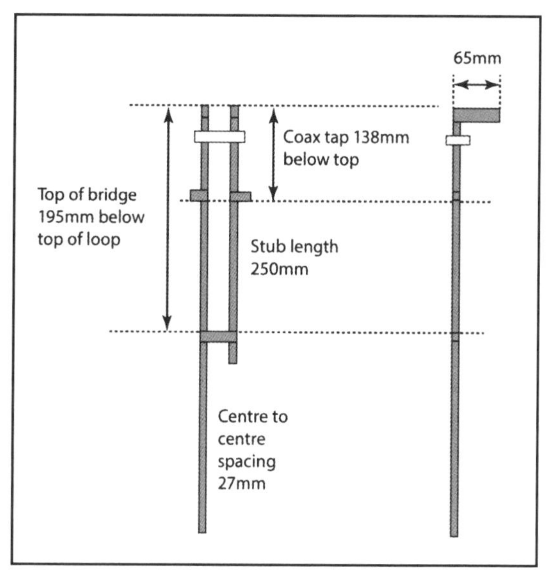

Fig 3.8: Essential dimensions. The insulator is shown in white; the position is not given as it is non-critical.

one side and one edge is curved over, which improves the rigidity and also locks with the edge of the support brackets to keep the loop horizontal. When you bend this material, do not make the 90° bends too sharp, or you may fracture the material.

The length of the support tube is optional, but to work properly the antenna should be mounted at least 0.34λ (0.7m) above the ground plane (indeed, the higher the better, though consider what overhead limits are likely to be encountered, for example in car parks).

On the principle that a picture is worth a thousand words, it should be possible to reproduce the antenna using the dimensions given in **Fig 3.7** and **Fig 3.8** plus the four photos on this and the next page.

Close up of the assembled Lambda Loop.

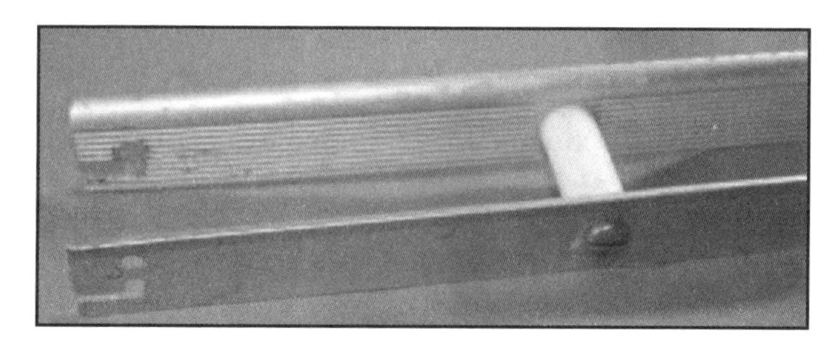

Detail of mating end of loop.

Three insulating pieces are used: two to hold the loop in shape and one to hold the stub tubes together at the open end. This latter was made from a piece of reinforced PTFE measuring 55 x 20 x 10mm. For the loop insulators two 30 x 12.5mm (1/2in) diameter PTFE rods were used; round has slightly less air resistance than square section, but there is no reason why they should not be made from the same rectangular material used for the stub support – hence one piece of material has been specified, from which all three insulators can be cut. Their position is not critical.

The two pieces used for mounting the loop and the shorting bridge are made by suitably bending the 20mm strips of aluminium around the tubes. The two tabs

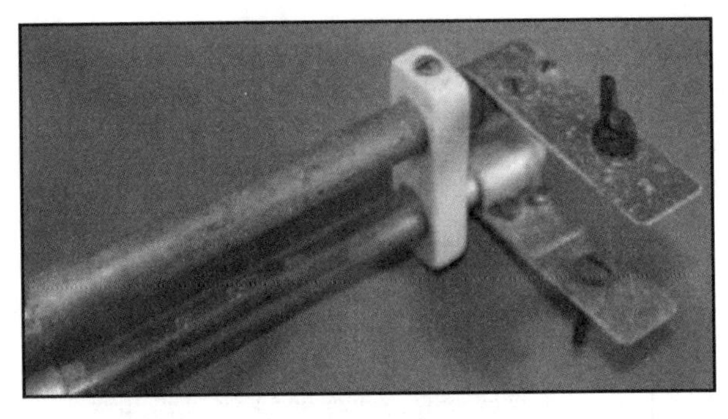

Detail of support brackets and feed.

Detail of loop connected to bracket.

holding the coax to the stub are made by cutting the fourth strip in half and bending these around the tubes. These are all secured by M3 nuts and bolts.

To feed the antenna, G3NAQ used a 1.5m length of RG-58 to reach the PA located in the boot. Although this coax is best avoided on the 2m band, over this length the loss is negligible provided there is a good match (SWR near to 1:1).

Finally, you will need something to hold it all together. G3NAQ used metric rust-less nuts and bolts, with toothed lock washers. For those with older stock, these could be substituted by BA sizes as indicated. You could also use self-tapping screws in most positions, but these might vibrate loose. Blind pop rivets would probably give a more professional finish, but assemble first with nuts and bolts until any adjustments that may be necessary have been made and only then replace them with rivets. However, bolts and wing nuts with lock washers are essential for fixing the loop to the support, for rapid but secure assembly and disassembly. A hammer was used to flatten the ends of these screws to prevent the wing nuts from getting lost.

The antenna was mounted with the feed lines pointing towards the front of the vehicle, as seen in the first photo. The loop is more stable in this orientation because it is supported forward of the air flow and so the loop streams naturally behind.

Although the notch in the polar diagram is in the direction one is travelling, this becomes less important with time if you are travelling towards the source. Also, the minimum radiation is in the direction of the occupants of the car, if you worry about such things (G3NAQ's car is a soft top, so no Faraday cage!) The actual method of mounting will depend on the vehicle and personal preference. In G3NAQ's case, he used a triple mag-mount on the boot lid, with a heavy-duty bayonet connector for rapid removal.

Adjustments were made to the stub with the antenna *in situ*, using a noise bridge with the mobile rig as detector. This has several advantages over adjusting for minimum SWR using a transmitter. First, no high level emissions are made while doing these adjustments, so there is no possibility of causing interference. Second, any level of mismatch is tolerated and, third, one can make the initial adjustments by ear, leaving the eyes and hands free for manipulation. Final adjustment is best done using the S-meter. However, starting from the dimensions given, very little mismatch should be found, so the use of a noise bridge should not be necessary. Adjustments to the stub length and tap point can be made by loosening the fixing nuts and sliding the tapping clamps

MOBILE OPERATION – IMPORTANT!

If you do operate while actually moving, be sure always to follow best 'hands-free' practice, always putting safe driving first. It is advisable to join an advanced driving institute (such as the UK's Institute of Advanced Motorists), receiving their guidance and taking their test, however long you have been driving – and however good you think you are.

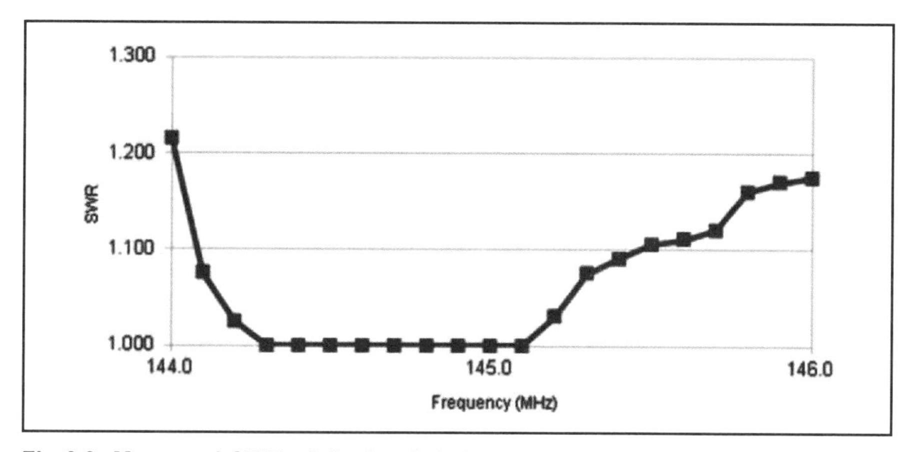

Fig 3.9: Measured SWR of the Lambda Loop. A Bird model 43 Wattmeter was used for these measurements, using a 100W 100 – 250MHz insert to measure forward power (just below 100W), and a 10W 100 – 250MHz insert to measure reverse power.

and / or shorting piece along the rods. Then, when satisfied, remove it and drill and tap for the M2.5 screws through the two mounting clamps, the shorting piece and the spacing insulator, for extra security and strength. You may like to finish it off and retain that 'new' look, as well as protect against the weather and particularly the salt used on the roads during the winter, by spraying it with a protective finish, excluding, of course, the contact surfaces between the loop and the mounting brackets (the photos on the previous pages were taken after the antenna had been in use for more than seven years).

It is never easy to measure completely the performance of an antenna. The SWR across the band is shown in **Fig 3.9** and is completely satisfactory. Neither the

polar plot nor the absolute gain have been checked, because it is unlikely that the measured values would be more accurate than the predicted values shown.

However, most amateurs will judge by the results. So, having completed the antenna, I drove up to a nearby high point, about 220m / 730ft ASL on the Berkshire Downs. Tuning across the beacon band only GB3VHF was strong, indicating that conditions were 'normal'. My first SSB CQ on 144.3 was answered by two stations, both using only 10W, and both stations were worked without difficulty. Joe, G0JJG, was located in Stowmarket, Suffolk, some 110 miles away, while Bernard, G0FIR, in Charlbury was only about 25 miles away, but it turned out he was using a fixed Yagi in the loft, pointing north (away from me). Subsequent use has confirmed its performance and reliability. I have worked using SSB to EI and DL from my area, the latter while actually mobile. The antenna has survived sustained speeds of up to 70MPH, and can be assembled or disassembled in about one minute.

COLLINEAR DELTA LOOPS

After all the small HF magnetic loops, and one small full-wave (VHF) loop, we end the chapter with a large HF loop antenna. The delta loop is a well-known antenna and can hardly be considered as 'novel' in its normal configuration. The 'ordinary' delta loop has sometimes been referred to as the 'poor man's quad loop', consisting as it does of a full-wave loop of wire, but folded into a triangular shape rather than the square shape of the quad.

One advantage of the delta loop over the quad loop, particularly on the lower-frequency bands, though, is that if configured with the apex at the top only one high support is required, as opposed to two for the square-shaped quad loop. If fed at, or close to, one of the lower corners as in **Fig 3.10(a)** the design provides vertical polarisation and a fairly low angle of radiation (about 20°), making it suitable as a DX antenna. However, if the delta loop is turned upside-down and fed at what is now one of the *top* corners, as in **Fig 3.10(b)**, it also provides vertical polarisation and a similarly low angle of radiation.

The delta loop in this configuration lends itself to being used in a novel two-element design. Two 'upside-down' delta loops can be fed together in phase to give a couple of decibels gain over a single loop (and perhaps 3 or 3.5dBd gain). A 'base up' delta loop can be constructed by having the top wire taut and allowing the other two 'sides' simply to hang down beneath it in a 'U' shape instead of the usual 'V'

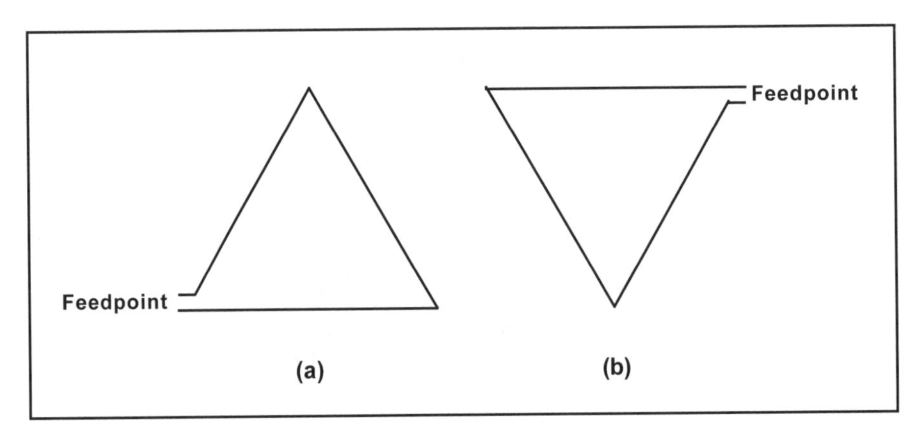

Fig 3.10: (a) Conventional vertically-polarised delta loop. An improved match can sometimes be achieved by feeding part of the way up the sloping side, at the point that is λ/4 from the apex. (b) The 'upside down' delta loop also provides vertical polarisation if fed at a top corner.

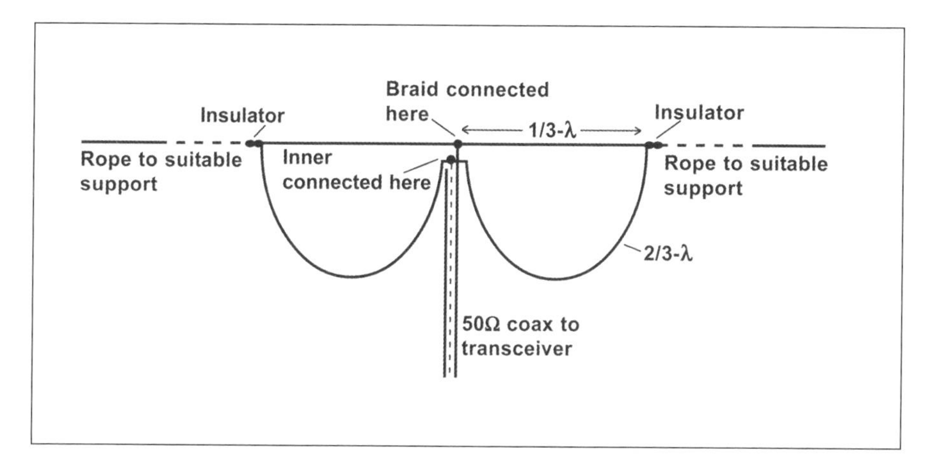

Fig 3.11: Close-spaced collinear delta loops.

shape of the delta loop. A second identical 'base up' delta loop is then placed next to the first, so that their two 'corners' are almost touching, and the two are fed together in phase. The resulting antenna is shown in **Fig 3.11**. In particularly windy locations it is advisable to anchor the lowest point of the two loops to the ground using an insulated cord.

In practice, instead of making two identical 1λ antennas, one single length of wire of 2λ long can be used. The suggested wire lengths for several HF bands are given in **Table 3.4**, although the antenna will probably need to be trimmed to length *in situ*. The impedance is around 75Ω and the SWR at resonance should be around 1.5:1 or better.

The top wire should ideally be horizontal but it may also be sloping if you only have one tall support and if that support is high enough to allow the lower of the two 'U' shapes to hang below the sloping wire and still have a reasonable clearance above the ground – at least 2.0m (6ft 6in) is suggested. I used this antenna in exactly this sloping configuration while on a work assignment in the 1990s in Papua New Guinea, from where I was active as P29DX. A 40m version was built, with the high end attached at the top of a 36m (120ft) professional tower. The far end was tied off to a distant tree with a long rope and the whole antenna system sloped from about 33m (110ft) down to about 21m (70ft). It could be argued that *any* antenna that high should work quite well, but it was compared with a 40m sloping dipole with the top also at 33m (110ft) high, and the collinear delta loops regularly provided around one S-point of gain in their favoured directions (broadside to the antenna).

Freq.	1/3-λ (half-top length)		Length of 'U' section		Total wire length	
7100	14.38m	47ft 3in	28.76m	94ft 4in	86.28m	283ft 1in
10125	10.08m	33ft 1in	20.17m	66ft 2in	60.51m	198ft 6in
14200	7.19m	23ft 7in	14.38m	47ft 2in	43.14m	141ft 7in
18140	5.63m	18ft 6in	11.25m	36ft 11in	33.77m	110ft 9in
21270	4.80m	15ft 9in	9.60m	31ft 6in	28.80m	94ft 6in
24940	4.09m	13ft 5in	8.19m	26ft 10in	24.56m	80ft 7in
28300	3.61m	11ft 10in	7.21m	23ft 8in	21.64m	71ft 0in

Table 3.4: Approximate dimensions for collinear delta loops in metric and imperial measurements.

Chapter 4
Beams

'WONDER BAR' (OR 'BOW TIE') BEAM FOR 17M

Gary Cook, W9JSN, was looking for an antenna that would be directional, rotatable and have some gain over a dipole. He also wanted it to be relatively small and inconspicuous, so as not to be the cause of any adverse comments from neighbours or family. In the January 2007 *QST* he writes that he remembered an article in *QST* dating from November 1956 which described a 10m 'Wonder Bar' antenna. W9JSN revived the design to make this compact beam antenna for 17m, simply by scaling the dimensions for the new band.

Each of the two elements are only 12ft 8in (3.86m) across, approximately half the length of the elements on an ordinary full-sized beam for 17m. The spacing is just 6ft 7in (2.01m). The whole antenna is built from materials readily available at hardware shops.

The two centre supports of the beam are cut from a single 15 x 20in (38 x 51cm) by 3/8in (3.2mm) thick Poly kitchen cutting board bought at a local household goods shop. A small power saw was used to cut across the cutting board 7in (18cm) from each of the ends on the long sides to obtain two 7 x 15in (18 x 38cm) pieces. Use an electric sander to smooth the edges and round the corners of the newly cut side of each piece.

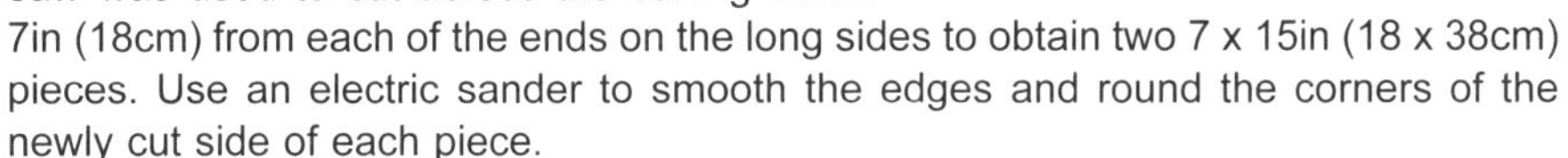

The eight diagonal spokes or 'radials' are each 77in (1.96m) long, and can be made from 3/4in (19mm) diameter aluminium or copper tubing. One end of each radial must be flattened for a length of about 2in (5cm) by compressing the tubing in a vice or by pounding them flat with a hammer on a cement block. Once flattened, drill a 1/4in (6mm) hole half an inch (12mm) from the end of each flattened section. The ends are fastened to the centre supports using 1/4in (6mm) hex bolts and 3/4in (18mm) U-bolts, as shown in **Fig 4.1**. Use washers and lock washers under all nuts to hold them in place and also to keep them from gouging into the cutting board centre support (you may also want to use extra washers and nuts as needed to make spacers between the centre support and the flattened ends of the radials). The two centre bolts for each element should be mounted 5in (12.7cm)

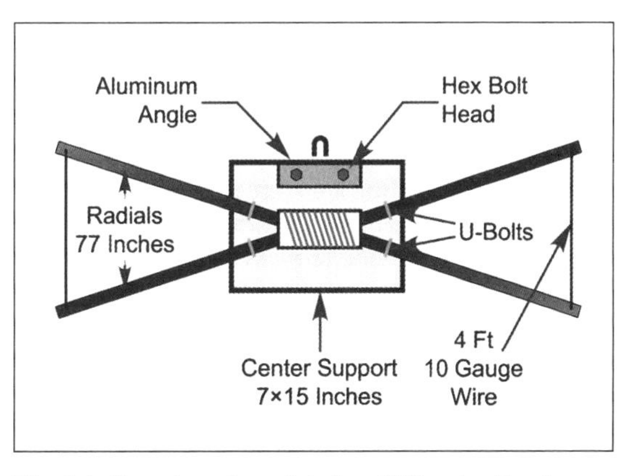

Fig 4.1: Construction details of Wonder Bar beam.

apart to allow room later for the centre loading coil. The radials should be fanned out by slightly more than 4ft (1.22m) at the far ends before they are fastened down with U-bolts. This will put a little tension on the wire tie bars once they are attached to the far ends of the radials. These are made of 10 gauge bare copper wire, slightly less than 4ft (1.22m) long, to maintain tension so as to hold them tight and to keep them looking straight.

There are identical loading coils at the centre of the element which are made from 7ft (2.13m) lengths of 12 gauge bare solid copper wire, wound on 6in (15cm) long sections of 1in (2.5cm) PVC plumbing pipe. Each coil occupies the centre 4in (10cm) of the PVC former. Drill small holes for the ends of the wire about 1in (2.5cm) from each end of the former. Insert one end of the wire far enough through the hole to allow several inches for eventual mounting and hook-up. Hold the other end of the wire in a vice to keep it tight; then turn the former slowly to wind the coil. Solder ring terminals to the wires extending from the ends of each coil, and fasten these to the two centre bolts using lock washers and nuts.

A smaller coupling coil was used to link the feedline inductively to the driven element of the antenna. This coil consists of two turns of 12 gauge insulated solid copper wire wound fairly tightly around the loading coil for the driven element. One end of this coupling coil is soldered to the inner conductor, and one end to the outer conductor, of a short length of 50Ω coax.

Fig 4.2: Close-up of the back side of the modified cutting board used as a centre support.

The boom and boom-to-mast mount were fashioned from as old CB antenna, but any suitable length and diameter of aluminium tubing would work just as well. And a small sheet of metal or heavy plastic (such as the leftover material from the kitchen cutting board mentioned above), along with two pairs of U-bolts (one pair for the boom and one pair for the mast), could be used to fashion a boom-to-mast mount. A 6in (15cm) length of 1.5in (4cm) aluminium angle stock and a U-bolt was used to mount each of the centre supports of the two elements to the boom (see **Fig 4.2**).

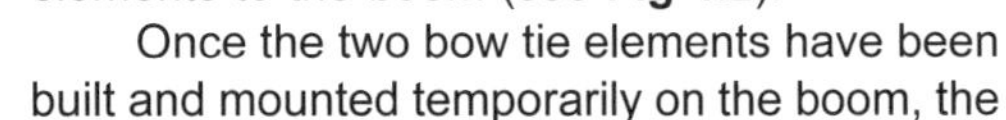

Once the two bow tie elements have been built and mounted temporarily on the boom, the only part of this project calling for critical measurements begins. Using an antenna analyser (e.g. MFJ-259), tune the centre loading coils of both elements to the proper frequencies, and then adjust the coupling coil on the driven element. Based on "a good deal of trial and error", W9JSN recommends the following procedure:

- Temporarily mount the antenna on a short mast or on top of a self supporting ladder where it will be well away from all metal objects and yet within reach for easy adjustment of the loading coils.
- Tune the coil on the reflector element, mounted by itself temporarily on the boom. Tune it so that it is resonant at a frequency approximately 5% below the bottom of the 17m band (around 17.165MHz). W9JSN removed one prong from a two pronged spade terminal and bent the remaining prong so that it would fit snugly on the wire of the loading coil but at the same time slide easily around the coil for purposes of adjustment. He soldered this spade terminal to an insulated 12 gauge stranded copper wire with an alligator clip at the other end. Once the spade terminal was set in place, he temporarily fastened the alligator clip to the wire at one end of the coil, thereby shorting out the turns between that end of the coil and the spade terminal.
- Begin by shorting out six or seven turns at one end of the coil. Then temporarily link the antenna analyser inductively to the coil and adjust the frequency of the analyser

until you see a small dip in the meter, indicating the resonant frequency of the element. The inductive link connecting the antenna analyser to the loading coil of the reflector can be a coupling coil of a two turns wound temporarily around the outside of the antenna loading coil, or a small coil inserted inside the coil former of the loading coil.

- The other end of the inductive link should be connected to the antenna analyser via a short length of coax and the appropriate plug. Once you determine the initial resonant frequency of the bow tie element you will know whether you need to short out more turns and thereby raise the resonant frequency, or the reverse. Keep adjusting the tap on the loading coil until you find the point at which the element resonates at the desired frequency.
- Use the same procedure to tune the driven element to the centre of the 17m band (about 18.12MHz). In this step the two turn coupling coil that will be a permanent part of the antenna can serve as your link between the loading coil and the antenna analyser.
- Finally, adjust the coupling coil on the driven element, by sliding it to various positions along the length of the loading coil, and also by varying the tightness around the outside of the loading coil, for minimum SWR. This is a crucial step and requires some experimentation. W9JSN found that he achieved the lowest SWR by locating the coupling coil near the end of the loading coil farthest from the shorted turns.

The readings you will get in the three final steps above are somewhat interdependent, so you may have to repeat the steps several times to find the best settings. It is likely that the resonant frequency of the driven element will shift upwards somewhat once the antenna is raised to its final position. By proper adjustment (and this definitely requires both persistence and patience), you should be able to obtain an SWR of less than 1.7:1 across the entire 17m band.

Once you find the best settings for the loading and coupling coils, solder all the connections with a heavy duty soldering gun. W9JSN used two 1/2in (1.5cm) plastic cable clamps fastened to the centre support of the driven element with small screws to hold the coupling coil, and the short length of coax attached to it, firmly in place. See **Fig 4.3**. This is important, since even a slight movement of this coil can affect the SWR of the antenna.

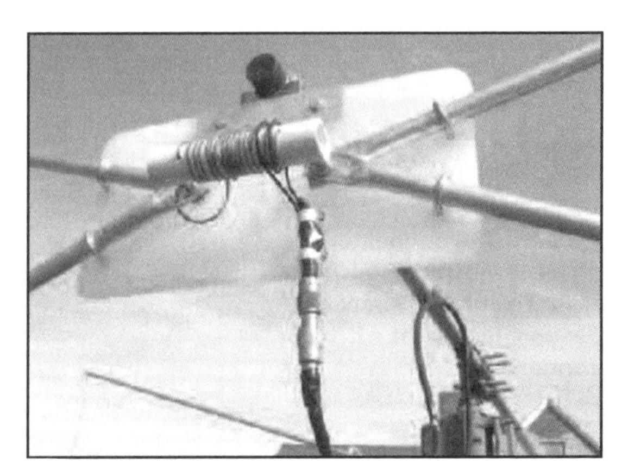

Fig 4.3: The front side of the centre support showing the loading coil, the coupled coil and the element radial mounting. Note the two plastic cable clamps that hold the coax and the coupling coil mechanically solid.

Casual tests run with other US stations indicate that the beam has a front-to-back ratio of approximately 10 to 15dB, and that both the gain and the directional pattern of the antenna resemble those you would expect from a full sized two element Yagi.

THE 'JUMPER BEAM'

Like many good antenna ideas, this one developed through necessity. In the 1980s Rudolf Klos, DK7PE, often stayed in high-rise hotels in Africa from where he wished to operate. The buildings proved ideal for supporting sloping dipoles for the lower-frequency bands, but results on the higher bands were often disappointing. The Jumper Beam was developed to address this issue. It was first described by DK7PE in the October 2011 *QST* and featured in the RSGB book *Successful Wire Antennas*.

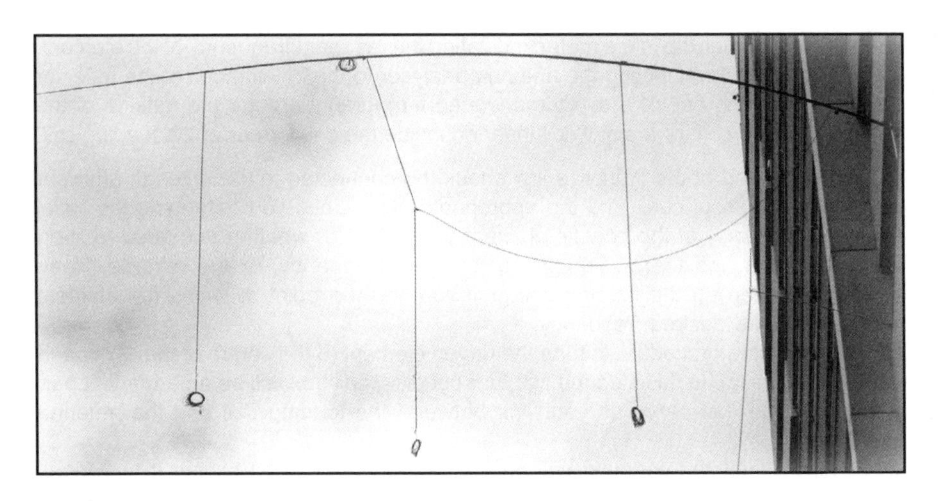

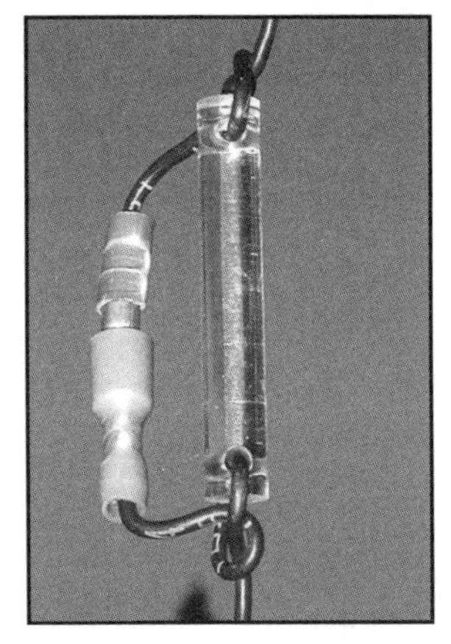

Detail of the jumper connection. The connector is an automotive push-on type. The plastic insulator keeps the weight off the connection.

DK7PE started by using a single vertical dipole supported by a long fibreglass pole and 'hung out' of the window of a high-rise hotel. Rudi wrote, "Fixing this vertical dipole in front of a big building [made] out of concrete and steel must have an advantage in [the] form of at least a little additional gain, as any rear energy that isn't absorbed must be reflected somewhere – even though the properties are hard to predict. So why not put a real, perfectly cut, reflector behind the vertical dipole and a director in front?" The result was the 'Jumper Beam': a full size wire beam that covers all bands between 40 and 10m. The correct wire lengths are connected by 'jumpers' (see photo) from the car industry for each band.

The antenna is suspended vertically from a 12-metre long fibreglass pole (e.g. a Spiderbeam pole – see www.spiderbeam.com). A 12m pole is long enough to build a three-element beam for all bands from 30m to 10m. On 40m the ideal spacing between the director and reflector is around 9m, making it too large, so a two-element Yagi with driven element and director is used on 40m. To change bands, simply open or close jumper connectors and wind up any spare antenna wire.

To keep the antenna as light as possible 20AWG copper wire is used. This is a good compromise between breaking strength and weight. The lengths that have been found to work best are shown in **Table 4.1**. No difference in performance was noticed when using a balun so, to keep weight

Band (m)	Director	Director Spacing	Driven Element	Reflector Spacing	Reflector
10	4.76	1.07	4.97	2.14	5.21
12	5.37	1.21	5.60	2.41	5.87
15	6.36	1.43	6.62	2.85	6.96
17	7.40	1.66	7.71	3.32	8.09
20	9.52	2.14	9.92	4.28	10.41
30	13.36	2.97	13.79	5.94	14.47
40	19.04	4.27	19.84	N/A	Not used

Table 4.1: Element dimensions and spacings.

Attaching the fibreglass pole to the balcony. Adjustable locking straps or 'bungee' cords are used to secure both the bottom of the pole and the fulcrum on the balcony rail.

to a minimum, the coax is connected directly to the dipole driven element without the use of a balun. Lightweight RG-58 coax is used: attenuation, even on the higher bands, is not an issue due to the short length required from the driven element to the transceiver and, with a properly matched antenna, RG-58 can easily handle 700W or more power.

The fibreglass pole must of course always be properly secured both on the balcony railing and at the inner end with a suitable counter-weight or locking strap. It is even possible to 'rotate' the beam to some extent by moving the inner end of the pole around.

Light winds are not a problem because the elements move with the wind, while in a strong wind the antenna can be dismantled in a few minutes.

Even though he has been unable to measure the gain, DK7PE reports that he has found the performance of this antenna to be far better than that of a single element antenna.

THE VERTICAL MOXON ARRAY

These days the Moxon Rectangle is a well-known design and no longer considered to be 'novel'. However, it is normally configured as a horizontal wire beam supported by fibreglass spreaders, as originally designed by the late Les Moxon, G6XN. Nigel Ramsey, G6SFP, has turned the Moxon Rectangle on its side to produce a novel 2-element vertical array. The wire beam is supported by two 10m fibreglass fishing poles. The structure is strengthened by fastening the two poles together where they cross over. Peter Millis, M3KXZ, also uses a vertical Moxon array close to the sea. He has described his version on a website dedicated to the Moxon Rectangle: see www.moxonantennaproject.com

G6SFP's vertical Moxon array.

M3KXZ gives the following dimensions for his 14MHz version of this interesting antenna:

Driven element:
7.26m vertical, 1.15m horizontal;

Reflector:
7.26m vertical, 1.42m horizontal;

Spacing
(between driven element and reflector ends): 0.11m;

Height of base above ground:
Variable, up to 1.00m.

As with all vertical antennas, locating the vertical Moxon array by the sea will give a great low-angle signal. By directing the array out to sea, the parasitic element will provide even more low-angle gain, while rejecting signals from the land side.

THE SPIDERBEAM

It would almost be a dereliction of duty to produce a book about novel antennas and not to include the Spiderbeam! Now thought of primarily as a commercial product manufactured by the company that took its name, the Spiderbeam antenna started off life as a construction project. The design described here was such a success that the Spiderbeam company was formed to manufacture the antennas commercially (and it also sells the well-known fibreglass Spiderpoles that can be used for making vertical antenna arrays). However, for those who want to build their own Spiderbeam a detailed construction guide is still available (in any of 24 languages) by going to the Spiderbeam website (www.spiderbeam.com) and clicking on "Documents".

The original Spiderbeam as described here covered 10, 15 and 20m and consisted of three interlaced wire Yagi antennas on a common fibreglass spider, with three elements on 20m and 15m and four elements on 10m.

Later, a 5-band version covering 10, 12, 15, 17 and 20m was made available. As a keen HF bands operator, I am using the commercially-made 5-band version at PJ4DX and, with the antenna no more than 10.5 metres above ground, worked 207 DXCC entities within the first three months of 2015. I can confirm that this antenna certainly performs!

The commercial 5-band Spiderbeam at PJ4DX.

The Spiderbeam designer, Cornelius Paul, DF4SA, described how the Spiderbeam came into being in an article in *RadCom*. His need was for a lightweight portable antenna covering the 10, 15 and 20m bands. In November 2003 DF4SA wrote, "Five years ago, the Spiderbeam was just a dream. I was not convinced by the exaggerated claims by the minibeam manufacturers of gain, front-to-back ratio and bandwidth. One day I stumbled on an antenna design called a 'bow-and-arrow-beam' (or 'Bird-Yagi' after its inventor Dick Bird, G4ZU). It is a three-element Yagi using director and reflector elements bent into a V-shape. Nowhere in the literature could I find a multi-band version to meet my requirements, so I decided to start development myself. After countless simulation runs, the Spiderbeam eventually evolved."

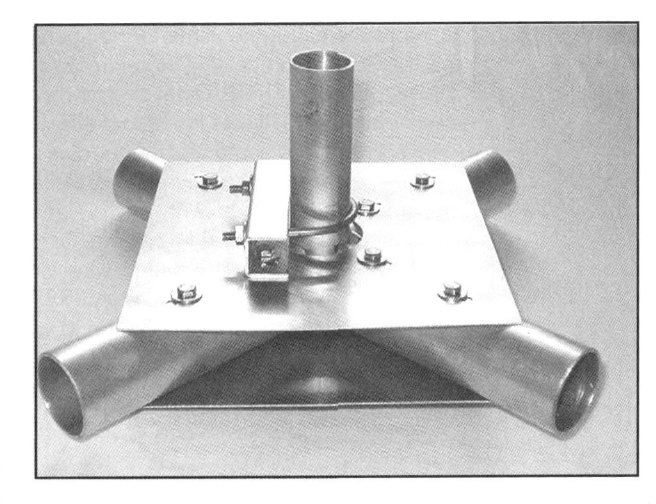

The central hub is made of aluminium tubes and plates.

The original Spiderbeam that DF4SA developed weighed only 5.5kg (just over 12lb) and, when dismantled, fitted into a pack only 1.20m long, making transportation easy. Only a light mast and rotator are necessary and the antenna has low wind resistance.

The heart of the construction is the central hub made from aluminium sheet metal and tubes (see photo above).

The fibreglass tubes that make up the boom and the spreaders are the bottom 5m lengths of 9m-long fishing rods. When inserted into the aluminium tubes in the central hub together they form the 'spider' that gives the antenna its name.

The spider gets its stability by being guyed completely from within, see **Fig 4.4**. The guys are Kevlar lines (1.5mm diameter, 150kg breaking strength).

Fig 4.4: The 'spider' that gives the antenna its name, before any of the wire elements are attached. The lines shown here are Kevlar guying lines, which give the basic structure its strength and stability.

The director and reflector elements are V-shaped, whereas the driven element is a multi-band fan dipole for 20, 15 and 10m, i.e. three individual dipoles connected at their centre feedpoint. The feedpoint impedance is 50Ω, fed through a W2DU type current choke balun. No phasing lines or matching devices are needed.

The lengths of the elements are shown in **Table 4.2**. The wire lengths specified are only valid for bare wire of 1mm diameter: other types of wire, and especially insulated wire, will require different element lengths because of the change of velocity factor caused by the insulation. The same is true when fixing insulators at the ends of the wire, as they also cause a change in the effective electrical length of the wire. It

Band (m)	Reflector	Driven element	Director 1	Director 2
20	10.54	2 x 5.02	9.84	–
15	7.00	2 x 3.47	6.48	–
10	5.26	2 x 2.62	4.88	4.88

Table 4.2: Spiderbeam element dimensions (in metres).

is important to cut the wires to the specified lengths: an error of just one centimetre will make a difference. It is therefore also necessary to use wire that does not stretch: the wire used by DF4SA was copper-clad steel wire (Copperweld®, DX-wire® 1.0mm, black enamelled). It was found that this wire had a unity velocity factor, i.e. the lengths derived from the computer model could be used directly in the real world. It also became clear that the covering of the element tips (4cm-long pieces of 8mm-OD polyamide hose, filled with epoxy) affected the resonant frequency of the wire elements; it drops by 100 to 200kHz. Of course, this effect must be taken into account when transferring the simulated wire lengths to reality! The specified wire lengths are a good compromise for CW and SSB operation. For single-mode operation it is, of course, very easy to use one set of wire elements optimised for CW and another one optimised for SSB, thus squeezing the last decibel out of the design.

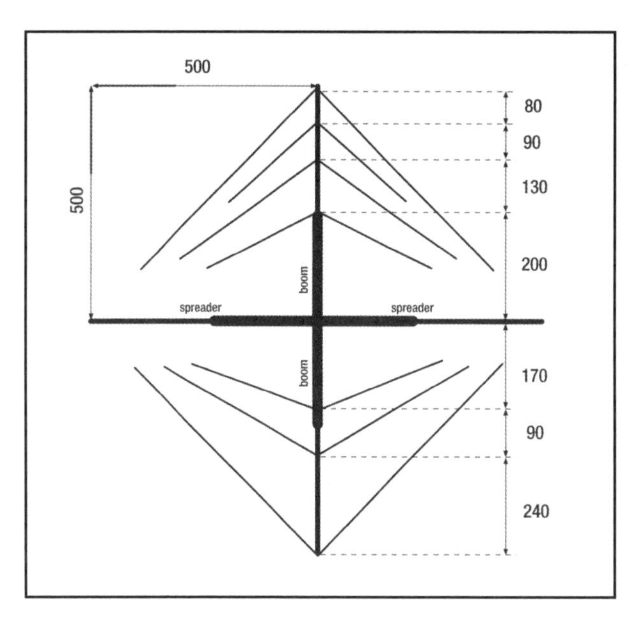

Fig 4.5: Placement of the parasitic elements. All the measurements are in cm.

The placement of the parasitic elements is shown in **Fig 4.5** and the three dipoles of the driven element in **Fig 4.6**. The three dipoles of the multi-band driven element must be correctly spaced vertically: the further apart they are spaced, the less is the mutual interaction, as with any multi-band dipole. The distance between the highest dipole (20m) and the lowest dipole (10m) should be around 50cm. It is also important to keep the 10m dipole a few centimetres away from the fibreglass spreaders, otherwise the VSWR will change a lot when the spreaders get wet from rain.

Full construction details can be found in the step-by-step Construction Guide available from Spiderbeam themselves.

The real-life performance of the Spiderbeam was found to be close to that predicted by computer simulation and is roughly the same as for a typical modern tribander with a 6m or 7m long boom.

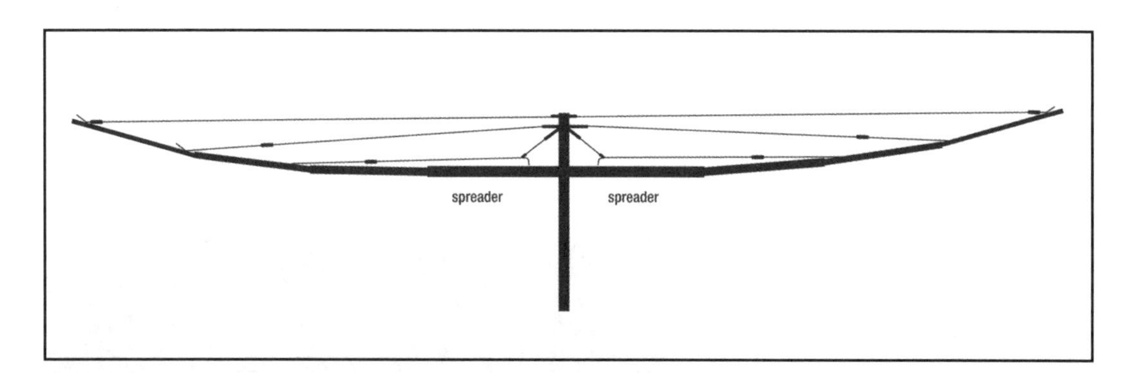

Fig 4.6: Placement of the three dipoles of the driven element on the fibreglass spreaders. Top: 20m; Middle: 15m; Bottom, 10m. There should be at least 50cm between the 20m and 10m dipoles and a clearance of several centimetres between the 10m dipole and the fibreglass spreaders.

THE GW3YDX 'SUPER MOXON'

In a book about novel antennas it can sometimes be difficult to know what to include and what to leave out. The Moxon Rectangle, designed by the late Les Moxon, G6XN, would definitely have been described as 'novel' at the time G6XN developed it. However, it proved to be such a successful design that by now it is regarded as something of a standard that is used by both home-brew antenna enthusiasts and commercial companies such as Optibeam.

The standard Moxon Rectangle therefore does not feature in this book. However, Ron Stone, GW3YDX – while being a keen enthusiast of the Moxon Rectangle in its basic form – discovered that the fairly broad frontal lobe of of 80° between the 3dB power points of a 6m version he built could be improved upon. The result is what GW3YDX calls the 'Super Moxon' – and *this* can certainly be described as a 'novel' antenna!

The original Moxon Rectangle has a driven element and a single reflector, with the ends folded back towards each other, thereby decreasing the wingspan of the beam considerably compared with that of a standard Yagi. Making the basic Moxon Rectangle is not difficult using the free software *Moxgen*, which allows input of wire gauge and frequency and results in a dimensional guide. *Moxgen* is downloadable from www.moxonantennaproject.com/design.htm and, most usefully, allows creation of a model in either *NEC* or *EZNEC* formats for further experimentation.

GW3YDX's idea was to add a pair of directors to the Moxon Rectangle in the form of a second rectangle, i.e. two directors with the ends bent back and joined with an insulator. The design and construction of the resulting antenna was described by GW3YDX in the July 2010 *RadCom*.

Several evenings were spent modelling the 'Super Moxon' using *4NEC2* and *EZNEC+*, after which a physical antenna was built. For the 50MHz version, according to the modelling, an increase in the boom length to just under 2m (just over double the original boom length) and with no increase in the wingspan, achieved an additional

A GW3YDX Super Moxon for the 4m band.

Band	A	B	C	D	E	F	G	H	J	K	L	M
6m	2160	395	280	105	290	310	60	2140	0	780	1201	1861
4m	1572	275	175	110	195	202	43	1572	0	560	860	1310
2m	730	135	86	55	82	90	12	730	0	276	434	615

Table 4.3: Tubing lengths for 6m, 4m and 2m versions of Fig X.X. All dimensions in mm, measured to tubing centres.

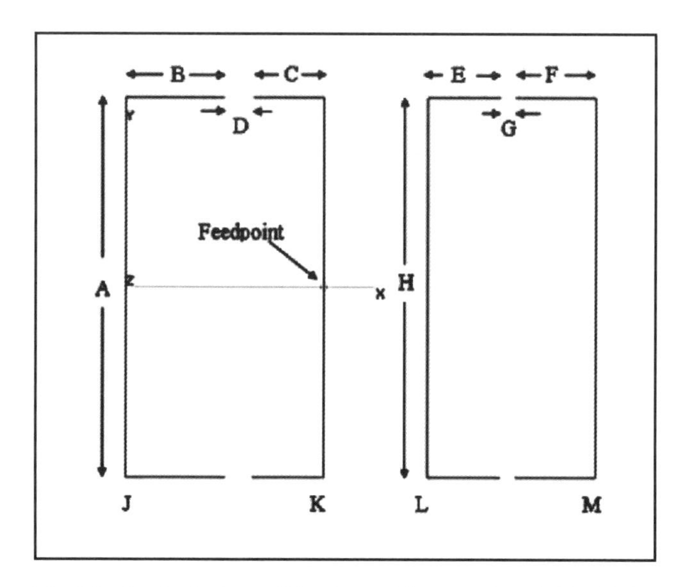

Fig 4.7: Basic design of the Super Moxon. See Table 4.3 for 6m, 4m and 2m version dimensions.

3dB gain with a 26.5dB front-to-back ratio and a VSWR of less than 1.5:1 between 50.0 and 50.3MHz when the model was optimised for 50.1MHz.

The antenna was built with aluminium alloy for the elements and fibreglass rod for the split driven element and the element-end insulators. This is a lightweight antenna, easy for one person to manage, and only a 1in (2.5cm) square boom was required. Construction of the antenna was principally with 0.5in (1.25cm) aluminium alloy, with the bent corners in 3/8in (1cm) material. 3/8in fibreglass rod was used for the element insulators both at the driven element centres and at the element ends.

Table 4.3 gives constructional sizes for 6m, 4m and 2m versions, the dimensions relating to **Fig 4.7**. The

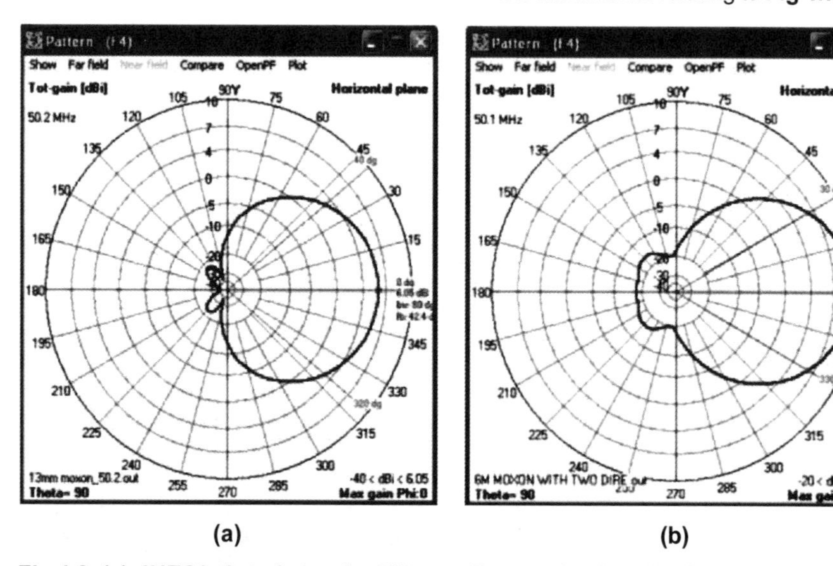

(a) (b)

Fig 4.8: (a) *4NEC2* plot of standard Moxon Rectangle showing fairly broad frontal lobe and 80° 3dB points; (b) *4NEC2* plot of Super Moxon radiation pattern. Note narrower frontal lobe and improved forward gain with very smooth rear pattern.

front part of the driven element is of course split at the centre and 50Ω feedline connected there through a balun. Although the final dimensions are as set out in the table, it is useful to slit the ends of the main element tubes and install stainless hose clamps for fine adjustment, particularly if other antennas are nearby.

The 6m antenna was tested at up to with 1000W without complaint. (The insulators on the directors on the 2m version are only 12mm long but at the 50W power level no arcing or instability was noted. Unfortunately no greater power was available on 2m for testing.)

The 2m version of the Super Moxon measures just 30in x 25in (76 x 63cm).

Fig 4.8 (a) and **(b)** illustrate the modelling results obtained. Note the decrease in –3dB power points to 60° compared with 80° for the original. This narrowing is where gain comes from. The front-to-back ratio of around 26dB is very respectable and the pattern has a nice clean 'light bulb' shape to it, with no minor lobes. Although modelling indicated that a 38Ω feed impedance was ideal, practical models have been fed with 50Ω cable and VSWR plots that follow the model graphs have been obtained with just slight adjustment of the driven cell element lengths.

The VSWR bandwidth is probably a little narrower than comparable Yagi designs, but at below 1.5:1 for the most-used part of 6m is very acceptable. GW3YDX comments (while admittedly tempting fate) that he has seen no 6m antenna design with comparable performance on such a short boom. A regular Yagi with the same gain would need a boom length of nearly 50% more and a turning radius nearly double of this design. A 10m version of this antenna would be the most space-saving means of achieving a good 'gain' antenna on the HF bands. "One can truly say that this design packs more dB into its size than anything else so far realised," GW3YDX said.

Note: *This design is being registered and therefore* ***commercial*** *manufacture is* ***not*** *permitted without permission of the design owner. However, radio amateurs may freely construct and use this antenna for their personal amateur stations. Commercial versions of the antenna are available from Vine Antennas Ltd (www.vinecom.co.uk).*

YS1AG MINI-BEAM FOR 40M

Those of us lucky enough to be able to erect a beam antenna for the 10 – 20m bands often baulk at the idea of a 40m beam, simply because of its size or weight. Even the so-called 'Shorty-Forty' commercially-made 2-element beams, using a loaded driven element and reflector, are bigger, heavier – and more expensive – than many of us are prepared to consider, as well as requiring a heavy-duty rotator and a large tower upon which to mount the antenna. As a result, most operators are prepared to forego gain and directivity on 40m for the simplicity of a dipole or vertical antenna.

A novel solution to this dilemma is offered by Andres (Andy) Goens, YS1AG, from El Salvador, in the form of a home-made mini-beam for 40m. Writing to Peter Dodd, G3LDO, Andy said that at first he had made a VK2ABQ 40m beam [similar to a Moxon Rectangle - *Ed*] but, "it had a very low Q and, with the first heavy rain, the wooden frame bent and it looked like a dead octopus. Now I have made a different one, a parasitic Yagi, which is much easier to tame". The resulting YS1AG 40m mini-beam was described in the January 2004 *RadCom*.

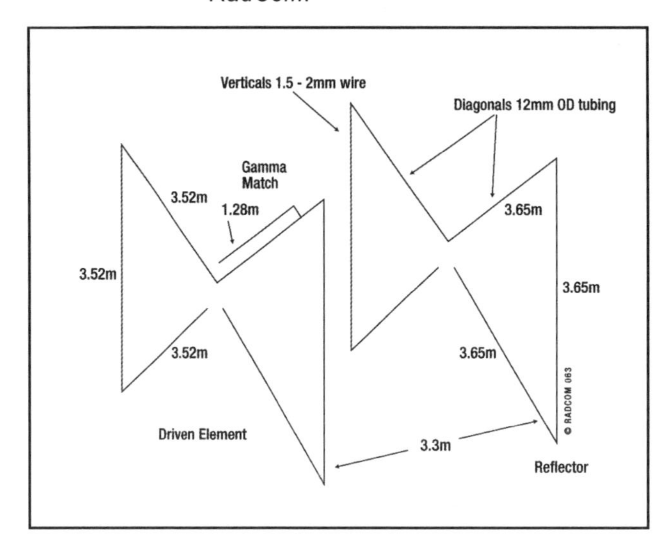

Fig 4.9: Dimension of the YS1AG 40m mini-beam.

The design achieves a small element size by folding the ends of the elements back on themselves. The antenna is shown in **Fig 4.9** and has a turning circle only one third the size of a conventional 40m beam.

Analysis of the antenna using *EZNEC3* indicates a maximum freespace gain of just over 4dBi and a front-to-back ratio of around 10dB, which agrees closely with the measured performance of the real antenna. Increasing the boom length from 0.078λ (3m / 9ft10in) to 0.1λ (4.26m / 14ft) would result in an increase in gain to 5dBi and a front-to-back ratio of around 12dB.

The feed impedance of such a small antenna is very low, and *EZNEC* indicates a value of around 4 to 5Ω. The feed arrangement used by YS1AG utilises a gamma match, in which the gamma rod is 1.3m long, made from 5 or 6mm OD tubing, and spaced 180mm from the driven element. The series compensating capacitor comprises two tubes, one sliding inside the other to make up a variable capacitor with a maximum value of 180pF. A 100pF doorknob capacitor is connected in parallel to make up the required total capacitance. The measured SWR, using such an arrangement with 50Ω feeder, is 1.8:1 at 7MHz, less than 1.2:1 over the range 7.04 to 7.1MHz and 1.8:1 at 7.2MHz.

G3LDO commented that waterproofing a Gamma-match variable capacitor can be difficult. Instead, he uses a temporary variable capacitor to make all the adjustments necessary to match the element to the feeder. He then removes the capacitor, measures the capacitance of the setting of the variable capacitor and replaces it with a fixed capacitor of the same value (two or more capacitors in parallel may be required in order to get the correct value). Fixed capacitors are normally inherently waterproof, although a coat of grease prevents degradation of the outside insulation of the capacitor due to prolonged exposure to our weather.

The centre supports for the antenna were constructed by welding four lengths of 20mm OD aluminium tubes to a centre ring made from a short length of larger diameter tube (the diameter selected to fit the boom). The construction is not unlike a quad 'spider' except that the angles are 60° / 120° rather than 90° / 90°. The 12mm diameter elements were fitted to the 20mm tube of the spider at the feedpoint end using reducers machined from aluminium bar. The high-voltage ends are insulated from the spider using hardwood dowelling and reinforced with fibreglass.

For those without access to aluminium welding equipment or a lathe, an alternative arrangement would be to use 1m lengths of aluminium angle stock. This material has two holes drilled in the centre to take a U-clamp, the size of which has been selected to fit the boom. Two lengths of angle material are clamped to one end of the boom so that they form a 60° / 120° spider. A further two lengths of angle are then clamped to the other end of the boom. When the angle sections are correctly aligned, the feedpoint end of the elements can then be fixed to the angle material with hose clamps. The high-voltage end of the elements can be supported using hardwood dowelling, which is fixed to the angle material. The wires forming the vertical part of the elements can be fixed to the ends of the 12mm elements using hose clamps.

But how does the mini-beam perform? YS1AG wrote, "On the air I have received consistent 2–3 S-unit reports over a local ham using a vertical. The front-to-back ratio is about 12dB. It is not the best antenna in the world, but it is much better than a dipole and very small indeed."

THE HEXBEAM

Like the Spiderbeam mentioned earlier in this chapter, the Hexbeam is a novel variation on the theme of a wire Yagi. Also like the Spiderbeam, the Hexbeam started off life as a 'home-brew' construction project, but commercial versions are now available from a number of different manufacturers.

But, while the Spiderbeam has (very nearly) full-size elements and it is the clever design that allows for a multi-element and multi-band antenna to be made with a relatively light weight, the Hexbeam has its wire elements bent into such a shape that the turning circle is only around half that of a conventional Yagi. The Hexbeam has two monoband elements per band and the novel design allows each band's elements to be 'nested' vertically above one another to give the correct spacing and thus provide monoband performance on each band.

It was found that the original Hexbeam design had a rather narrow bandwidth until further development by Steve Hunt, G3TXQ, resulted in the 'G3TXQ Broadband Hexbeam'. It is this design that is available commercially in the UK from Ant David, MW0JZE, who – with G3TXQ's permission – hand makes the antennas to order: see MW0JZE's website at www.g3txq-hexbeam.com (Most commercial Hexbeams cover the five bands from 10 to 20m, though the MW0JZE version also has two elements on 6m and the latest version [in 2015] adds a single dipole element bent into a square shape for either the 30 or 40m band, to provide no fewer than seven bands from a single antenna.)

The Hexbeam can, of course, still be home made and its design was described in the January and March 2008 'Antennas' columns in *RadCom*. This was around the time when G3TXQ was carrying out his redesign work to make the antenna more broad-banded. He wrote: "A consequence of the [classic] Hexbeam's geometry is a relatively narrow performance bandwidth; typically the F/B exceeds 10dB over a band equivalent to only 1.4% of the centre frequency, and the SWR is above 2:1 for a significant proportion of this band. This narrow bandwidth is largely determined by the Q of the reflector, which I measured at about 30 for a 10m element constructed from 16SWG wire. Compare this with a linear dipole, which has a Q of about 10. If we can find a way of reducing this Q we should end up with a broader-band antenna.

"I spent many hours modelling reflectors and evaluating ideas on a 10m test-bed... I tried using thicker wire of various types, including two varieties of coaxial cable and 'caged' wires. I also tested alternative reflector shapes. Of all the ideas evaluated, by far the most effective and easiest to implement was to change the shape of the reflector as shown in **Fig 4.10**. Even

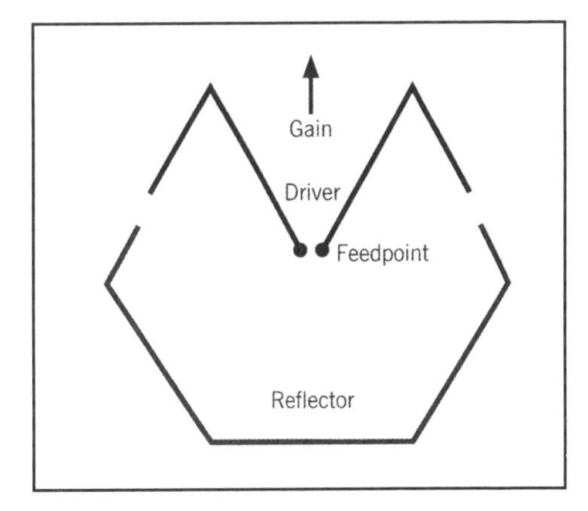

Fig 4.10: Revised element shapes of the improved G3TXQ Broadband Hexeam (see text).

Band	20m	17m	15m	12m	10m
Driver (half-length)	218	169.5	144.5	121.7	106.8
Reflector (total)	412	321	274.4	232	204.4
End spacing	24	18.5	16	13.5	12
Vertical spacing from 10m elements	38	15	9	5	0

Table 4.4: G3TXQ Broadband Hexbeam dimensions. (All dimensions are in inches: to convert inches to cm multiply by 2.54).

when using relatively thin 16SWG wire this shape has a radiation resistance of 44Ω and a Q of about 17. It requires an increase in turning radius of about 15%. Modelling a Hexbeam with this geometry produced very encouraging results with a F/B >10dB and SWR <2:1 across all of the 20m, 17m, 15m and 12m bands, and approximately 1MHz of the 10m band (listening tests with DX stations indicate that the F/B could be as high as 30dB). The modelling suggested there was little to be gained by making the same change to the shape of the driver element; in fact, retaining the classic shape for the driver delivers a better match to 50Ω and avoids a further increase in the turning radius.

"Construction and testing of a 10m monoband version of the new antenna confirmed the modelling results, and so a full 5-band test beam was constructed. The 20m, 17m and 15m results were immediately satisfactory, but it took some time to optimise the 12m and 10m performance; the proximity of these bands often causes problematic interactions which are not always predicted by the modelling, and the final wire dimensions for these bands were a result of 'cut and try' on the test-bed.

"The final dimensions using 16SWG bare copper wire are shown in **Table 4.4**. The band feedpoints are interconnected with 50Ω coax, and the array is top-fed as shown in **Fig 4.11** and the photo below.

"This 5-band design requires a horizontal distance of about 130 inches [3.30m] from the centre post to the tips of the spreaders. If you are unable to accommodate this increased size, *don't* be tempted to stick to the classic shape for 20m and adopt the new shape for 17m through 10m: modelling shows that the 20m performance bandwidth suffers dramatically. This is probably due to the mid section of the 17m reflector providing an RF coupling path between the 'knees' of the 20m reflector."

Steve Hunt, G3TXQ, earlier wrote, "Leo Shoemaker, K4KIO, and I have cooperated on the content of our respective websites – we agreed that I would major on the theory and experimental work, while he covered practical constructional information." The very detailed

Fig 4.11: Feed line wiring configuration for the five-band Hexbeam. 50Ω coax is connected to the *top* of the Hexbeam at the terminals for the 20m band. Be sure *not* to transpose the connections to each driven element.

Feed method for the multiband Hexbeam by K4KIO.

The completed five-band G3TXQ Hexbeam.

description of how to construct a five-band Hexbeam on K4KIO's website includes many photos, with appropriate captions, shown at every stage of construction. The main construction task is building a centre hub to support the six fibreglass element support poles. This must also have provision for fixing an insulated vertical support for the multiband feed system and of course provision for mounting on a mast. Note that the spacing between the ends of the reflector and the driven element ends is critical when it comes to gain and front-to-back ratio performance.

Both the G3TXQ and K4KIO websites are essential viewing for anyone interested in building a Hexbeam. They are at www.karinya.net/g3txq/hexbeam and www.hexbeam.com respectively. It should be pointed out that K4KIO also offers commercially-made Hexbeams, as do several other manufacturers in North America and Europe.

The Hexbeam provides an excellent solution for a beam antenna for those with restricted sites. The free-space gain (5.7dBi) and F/B (24dB) figures predicted by the computer model are really quite astonishing for such a relatively small antenna.

THE 'MINI HORSE' (MH) ANTENNA

This antenna may look a little like the Moxon Rectangle but is quite different. While the Moxon only has two elements, this is a three-element mini-beam. It was designed by Martin Hedman, SM0DTK, and described by him in the March 2010 *QST*.

Why is it called the 'Mini Horse'? Because it kicks like a grown-up horse but is as small as a pony!

The configuration of the Mini Horse (MH) antenna is shown in **Fig 4.12**. Note that it is basically a three element Yagi that is shortened by end-loading the driven element and folding back the parasitic elements along the diagonal support lines. The dimensions for various bands are given in **Table 4.5**.

The design can be made either as a rigid structure, suitable for mounting on a mast and rotating, or as an end-supported flexible wire array that can be suspended between two fixed supports.

MHz	A	B	C	D	E	F	G	H	L
144	1' 2.2"	0' 4.7"	0' 3.5"	0' 1.8"	0' 7.1"	0' 4.7"	0' 6.5"	1' 4.5"	1' 4.3"
28	5' 11.6"	1' 10.8"	1' 6.9"	0' 9.4"	2' 11.8"	1' 10.8"	2' 8.3"	6' 11.4"	6' 11.0"
21	7' 11.6"	2' 6.3"	2' 1.2"	1' 0.6"	3' 11.6"	2' 6.3"	3' 6.9"	9' 3.0"	9' 2.6"
14	11' 11.3"	3' 9.7"	3' 1.8"	1' 6.9"	5' 11.2"	3' 9.7"	5' 4.6"	13' 10.9"	13' 9.7"
7.1	23' 10.5"	7' 7.3"	6' 3.6"	3' 1.8"	11' 10.5"	7' 7.3"	10' 9.1"	27' 9.8"	27' 7.4"

Table 4.5: Dimensions of the Mini Horse antenna. The letters refer to the dimensions in Fig 4.12.

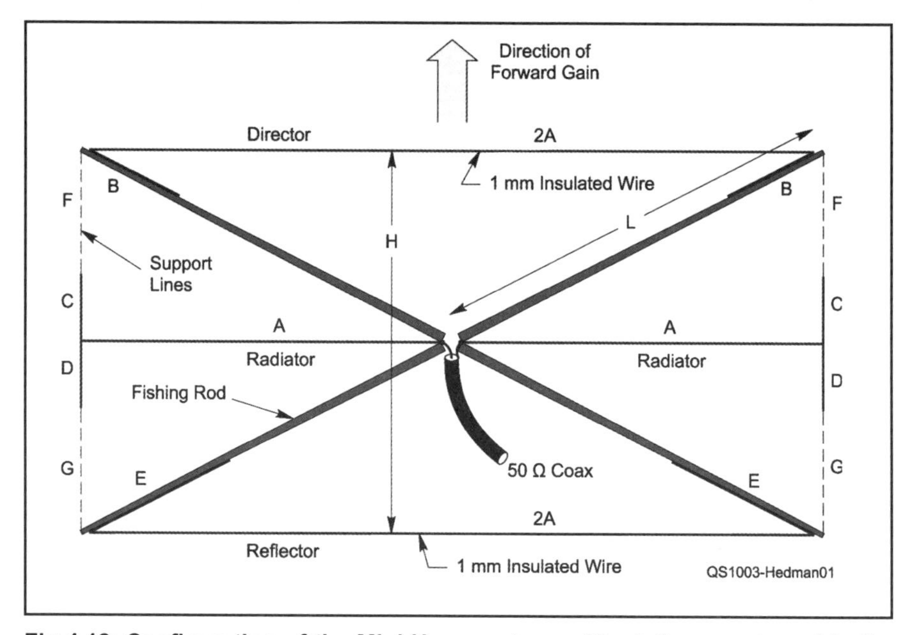

Fig 4.12: Configuration of the Mini Horse antenna. The letters are keyed to the dimensions in Table 4.5.

The driven element has its ends formed as a T with the longer tail ('C' in **Fig 4.12**) pointed towards the director and the shorter tail ('D') towards the reflector. The ends of the director and the reflector are bent and fixed to the spreaders (made of fishing poles) using electric tape. Support lines are fixed to the ends of the spreaders and these lines carry the ends (the 'T') of the driven element.

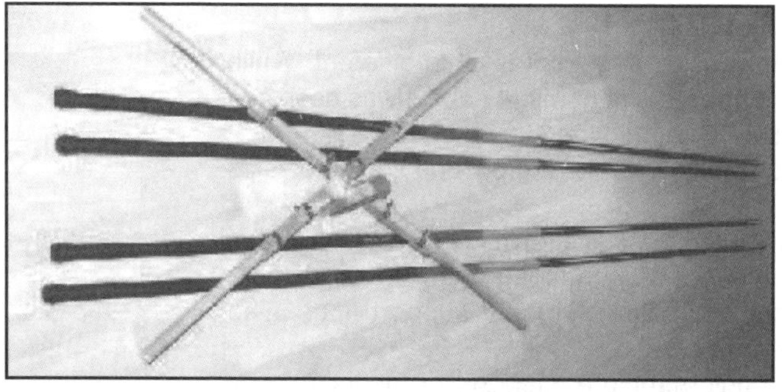

The aluminium angle stock pieces, bamboo poles and spreaders used for the rotatable version.

The spreaders are slid into bamboo poles that are fixed to a cross brace of aluminium angle stock using hose clamps. Another two pieces of aluminium angle stock are fixed to the cross brace to attach to the mast as shown in the photo to the left.

The antenna can also be constructed as a fixed wire array suspended between two fixed supports such as trees or masts. Although the wire dimensions of either version are the same, the fixed

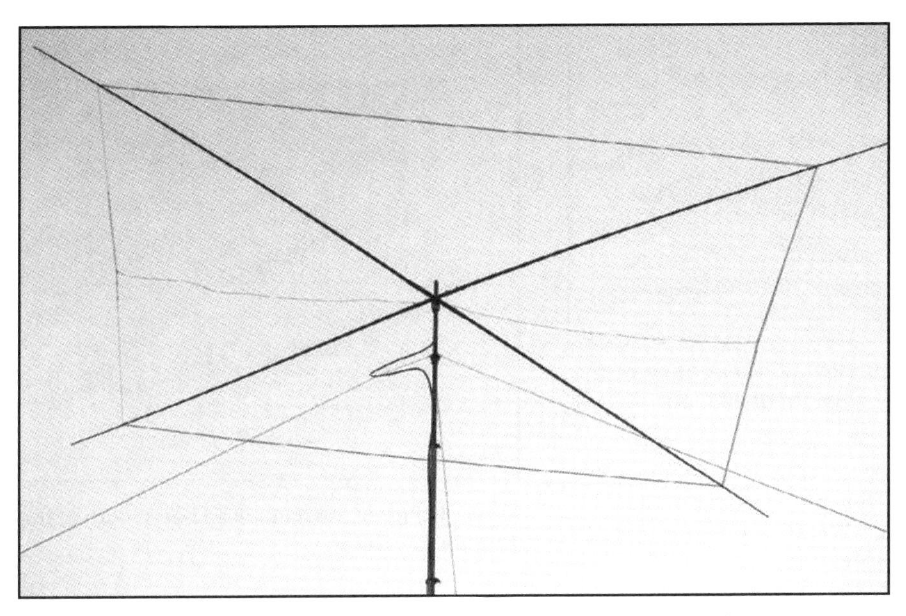

The 21MHz rotatable MH antenna, mounted here at a height of 30ft.

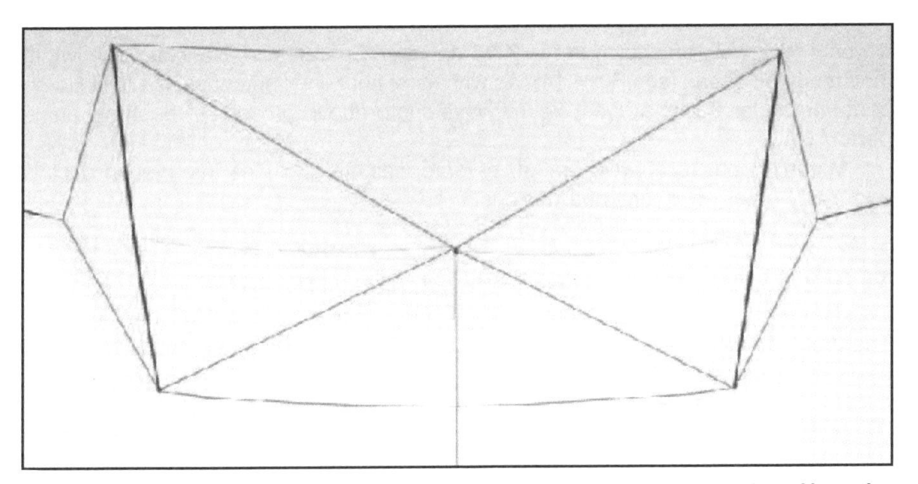

Alternative method of construction. This 20m MH is the fixed version. Note the difference in the position of the pole supports.

version is made using two 14ft (4.27m) poles, one at each end of the array as shown above.

Both designs use a small plastic box to hold the coaxial connector and the screws with which to connect the driven element, as shown in the photo. A slice of PVC tubing is attached to the box to facilitate assembly to the mast using hose clamps. In the fixed configuration, the diagonal support lines are passed through holes at one of the ends of the box.

SM0DTK used *4NEC2* for modelling this antenna but commented that it predicted a resonance higher in frequency than his physical design using the same

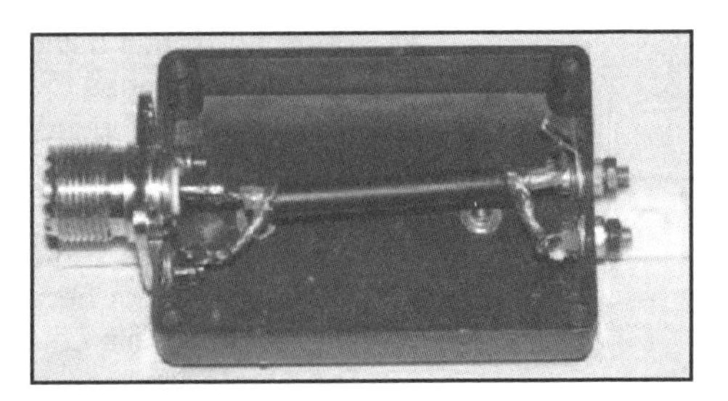

A plastic utility box provides a convenient connection arrangement by using it to mount an SO-239 coax socket.

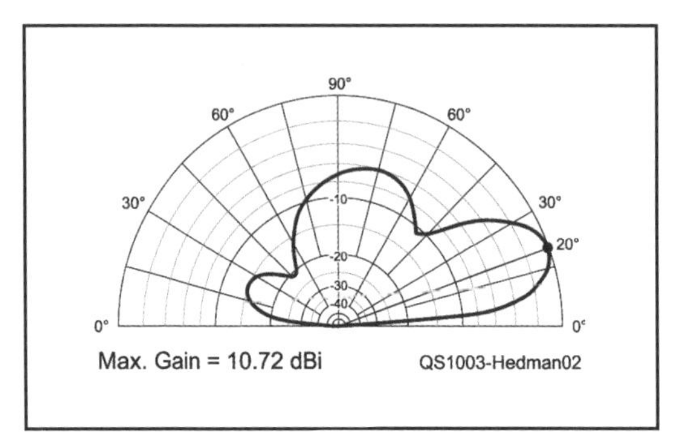

Fig 4.13: Modelled elevation response of the 15m Mini Horse at a height of 30 feet.

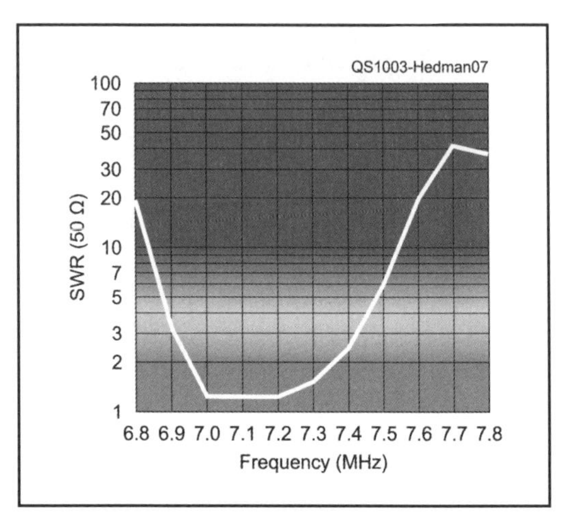

Fig 4.14: Plot of the measured bandwidth of the 40m fixed version.

dimensions. For example, his 15m antenna resonates on 21.2MHz while the *4NEC2* software predicted a resonance at 22.6MHz. The same factor (1.07) applies to all bands and might be the result of the use of insulated wire or the partly folded elements.

SM0DTK has built this antenna for the 2m band and for a number of HF bands. It works very well, as shown in **Fig 4.13** and also operates over a wider bandwidth than many antennas (see **Fig 4.14**). As with most horizontal antennas, the Mini Horse performs better if it is at least half a wavelength above ground for best low angle performance.

Martin's website at www.sm0dtk.se/antennas.htm includes information on this and many other novel antenna designs.

THE 2-ELEMENT PENTAGON BEAM FOR 40M

Quads are fine antennas and, on the higher frequencies, are relatively easy to build. The problem with quads on the lower frequency bands such as 40 and 80m is their size – they are physically huge. Writing in the April 2004 *RadCom*, Bruce Fleming, KI7VR, described a novel way to get a low band 'quad' in the air using a small tower.

In free space, the ideal loop antenna comprises a *circular* conductor with a circumference equal to one wavelength, or multiples thereof. Circular loops are indeed used on the VHF and UHF bands, but on the HF bands, a circular shape is hard to achieve and loop antennas are usually square or diamond shaped (quad loops) or triangular (delta loops). The conventional square quad needs support at *two* high points per loop – normally achieved with long spreader poles mounted at 45° to the horizontal using a special bracket on the boom (lots of torque where the spreader pole attaches to the bracket) – this is difficult to engineer for 40m and 80m. The fixed 'diamond' quad (**Fig 4.15(a)**) is simpler and more robust, with its single vertical spreader per loop – but there is a drawback: the diamond quad needs 1.4 times more vertical height than the square quad. The delta loop (**Fig 4.15(b)**), with its flat base, needs less vertical height than the diamond quad; however this 'less open' type of loop may not be as efficient as the square-shaped loops (John Devoldere, ON4UN, writing in his *Low-Band DXing*, calls the delta loop the "poor man's quad").

The Pentagon loop is more 'open' and requires less height than a diamond quad, yet still hangs from just one corner. A diamond quad for 40m has sides 10.87m long and a total circumference of 43.5m. When suspended from a corner, it occupies a vertical height of 15.4m. The corresponding delta loop has sides 14.5m long and

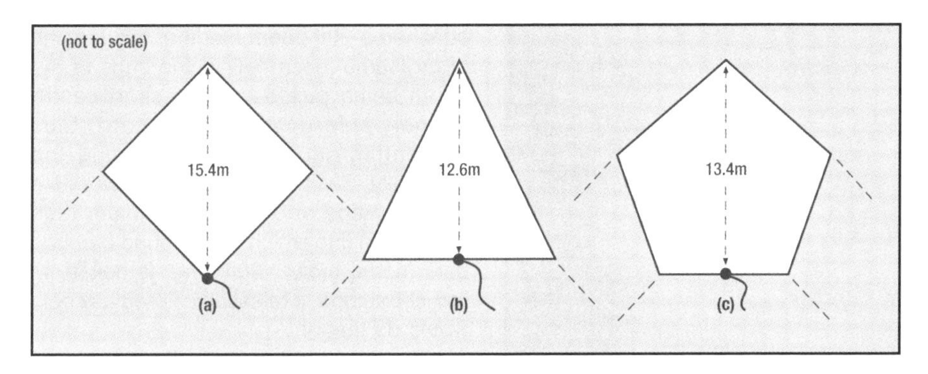

Fig 4.15: A comparison of three types of loop antenna.

a height of 12.6m. The Pentagon loop (**Fig 4.15(c)**) has sides 8.7m long and a height of 13.4m. The Pentagon requires only 87% of the vertical height needed by the diamond quad.

KI7VR commented that he found that by making the elements into a Pentagon shape a 40m loop antenna could be built without too much 'heavy engineering' being involved. The compromise is in regard to rotation: don't try to turn it! A fixed loop is much simpler to build. Obviously a non-rotating, 2-element loop should be aligned for the most important DX direction, but it can also be readily switched 180° by changing the parasitic element from a reflector to a director.

The Pentagon loop requires but a single vertical support at the apex. The two upper corners are pulled out with nylon cords to open up the loop. The bottom is made flat, as with a Delta loop, to minimise the vertical height requirement. It is apparent that the Pentagon loop is more 'open' than either a triangular or square loop and thus might be expected to radiate and receive somewhat better. A Pentagon has a 10% greater capture area than a square antenna made with the same wire. Tensioning the cords at the lower corners raises the bottom wire a few feet. The bottom ropes are probably not really necessary – without them, the wire droops in a catenary curve and although the wire hangs closer to the ground, the loop opens up even more.

KI7VR was given an old 30ft crank-up tower which formed the basis of the Pentagon beam. On top of the tower he mounted a 10ft stub mast made of steel fence railing (1.375in diameter). To the stub mast (8ft above the top of the tower), he

DIMENSIONS OF THE 2-ELEMENT 40m PENTAGON

Design frequency:	7050kHz
Driven element:	Side lengths: five equal sides of 8.7m Total length: 43.5m.
Reflector length:	103.5% of driven element: four sides of 8.7m, bottom side: 10.2m. Total length: 45.0m.
Note 1:	The resonance shifts a bit with soil moisture content; the loops needed to be shortened as the soil dried out during summer.
Note 2:	If the parasitic element is to be a director, its length needs to be 97% of the driven element, i.e. 42.2m, e.g. four sides of 8.7m and a bottom side of 7.4m.

The KI7VR 40m 2-element Pentagon beam. (The wires have been re-touched to make them more visible.)

attached a 20ft boom made of two lengths of the same fence railing, the ends of the boom being pulled up to a short centre support pole using Dacron cord, thus preventing sag and strengthening the assembly. Each end of the boom carries a vertical piece of 3in OD aluminium tube (18in long) into which a varnished bamboo pole (home grown, 20ft long) is bolted. In lieu of bamboo, a telescopic fibreglass pole would do the job.

The height at the top of the bamboo is 58ft. The antenna wire (copper-clad steel, multi-strand 14-gauge) is attached to the top of each bamboo pole. A small loop is soldered into each wire at the four remaining corners and Dacron cord attached to each corner point. How you pull out those top corners depends on your property. If you have lots of space (e.g. a field) you can extend the Dacron cords far out until they approach ground level and then attach them to ground stakes. In KI7VR's case, he had to climb nearby trees and put the Dacron cords through smooth plastic rings fixed to convenient upper branches.

After final adjustment of the cords, the bottom wire hangs about 19ft above the ground. By loosening the ropes, it is possible to drop the bottoms of the loops low enough so that, using step ladders, the coax connection can be modified, or the length of the parasitic element changed.

The two-element Pentagon described has an input impedance of about 100Ω when fed in the centre of the lower side. For matching, KI7VR used a quarter-wave transformer consisting of 23ft of 75Ω RG-6/U coax attached at the bottom centre of the driven loop. This connects to RG-58/U (50Ω) running to the shack. The VSWR at the shack is 1.2:1 at resonance. RG-213 would be better than RG-58/U.

KI7VR's location is on steeply sloping ground to the south so that, with the antenna pointed at ZL, the boom has an effective height of 175ft. With 400W into the antenna on 7002kHz, the reports from ZL and VK are that "it sounds like a local". Received signals are up by about two S-points over a dipole and the noise is the same or lower. Long path results to Europe have been excellent with the antenna pointed at New Zealand.

THE SKELETON BEAM

This clever design by Tony Preedy, G3LNP, is a rotatable derivative of the old W8JK antenna and covers all the amateur and CB spectrum from 10.1 to 29.7MHz. It is probably the best value and easiest to build 6 or 7 band HF rotary beam antenna that you will find! It was first described in the January 2014 *RadCom*.

Two horizontal dipoles driven with currents of equal amplitude, but 180° out of phase, can make an easy to construct bi-directional beam antenna, as proposed many years ago by W8JK. At a low height it has more gain at low take-off angles than when the same dipoles are operated as a parasitic array. In its simplest form the antenna has its elements suspended on spreaders between a pair of masts. Up to 3dB more gain may be achieved if the dipoles are extended, but the beam

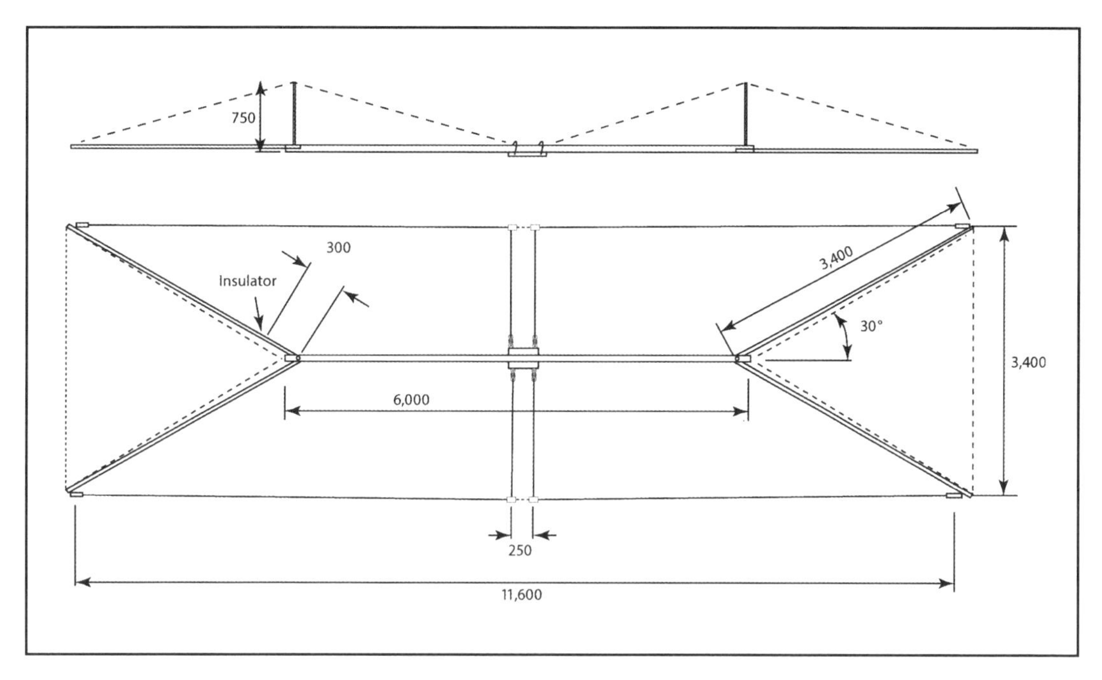

Fig 4.16: Dimensions of the Skeleton antenna. All dimensions in mm.

then gets very narrow and unless the antenna can be rotated your contacts may be limited.

Fig 4.16 shows the essential dimensions of the antenna, which is shown aloft in the photo to the right. The skeleton frame serves only to support the elements and allow their rotation. The central spine has no induced current and does not therefore contribute to or distort the radiation pattern. An element length of 11.6m was chosen because at more than a wavelength a pair of minor lobes are produced each side of the main beams. With this length these are better than –20dB down at 29.7MHz. Element impedances reduce with close spacing and ultimately will reach the point where the resistive component is comparable with losses in the wires. With spacing of 3.4m and thickness of 5mm the loss of gain is less than 1dB at 10.1MHz. Elements are formed from RG59 coaxial cable to achieve the necessary wire thickness without excessive weight. The inner steel conductor and outer copper braid are joined and sealed at each end.

The elements are not resonant at any working frequency. To allow the wide spectrum to be covered efficiently, phasing lines need to be short and of the highest characteristic impedance. They are therefore made of 1mm wire with 250mm separation and reach only to the centre of the array. Further separation has an insignificant influence on impedance, a characteristic that allows relative movement with negligible impedance fluctuation in windy conditions. The resulting impedance

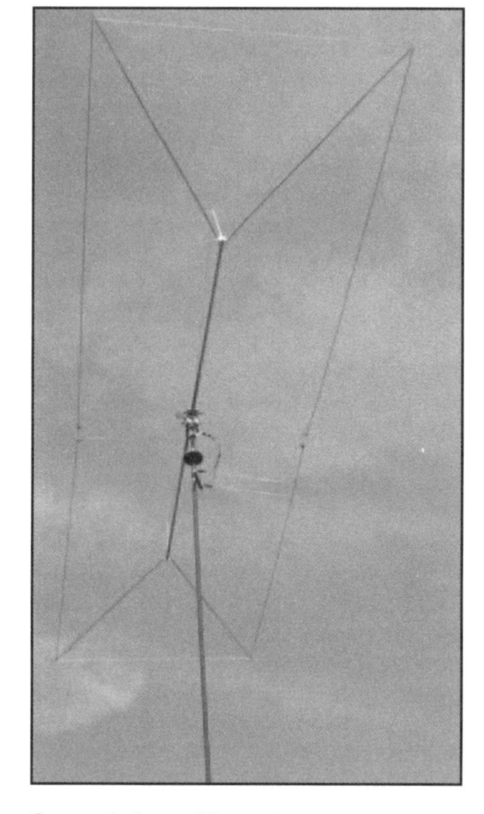

General view of the antenna.

Close up of the RG59 element ends and supports. The inner conductor and copper braid are joined and sealed at each end.

of 720Ω is also used for the main feeder because the wide separation allows an unsupported run of about 50m without spacers between mast and shack. Standard 450Ω ribbon feeder may also be used.

Unless a remote balanced tuner is attached to the antenna there is no possibility of using a main feeder of coaxial cable because the high SWR would result in low efficiency. However, to allow for rotation, a 1.5m length of 450Ω ribbon feeder is used to connect the phasing lines to the main feeder. The main feeder connects via a balun, described later, outside the shack and then via a few metres of coax to an automatic tuner. G3LNP used 1mm soft enamelled copper feeder wires that had been hardened by stretching them to 101 or 102% of their original length. The feeder wires have equal tension of 10kg provided by a pulley and weight system. Insulators for the elements are fabricated from sheet insulation material and they grip the RG59 by looping it through a pair of holes as seen in the photo to the left. At the element centres a length of Dacron sets the spacing between a pair of insulators.

The braced skeleton frame is constructed from aluminium tubes. The spine is made from two 2m lengths of 2in OD 16 SWG with a centre 2m section of 2in OD 10 SWG joined by 30cm sleeves of 2in ID 16 SWG. At each end a pair of 16 SWG aluminium arms are attached that taper from ¾in to ½in diameter. These were salvaged from an old Yagi antenna but except for the first 30cm may be bamboo or glass fibre. To prevent them from filling in the nulls off the side of the radiation pattern they have an insulator of Tufnol rod 30cm from their thick end. The arms are fixed to the spine with a 6mm bolt and held at 30° to the spine by shoulders of 32mm x 5mm aluminium bar (see photos below). A spacer of 3/8in tube is fixed inside to prevent the bolt from crushing the spine. This is installed before fixing the

Above: How the arms are braced from the spine.

Left: Preparing the end of the spine.

strut or forming the tabs, to which the shoulders are attached with self tapping screws. Please be aware that some grades of aluminium, sold for masts, may not be sufficiently malleable to allow the tabs to be bent without fracture. In this case an alternative is to fit a disc of 8mm material inside the spine.

At each end of the spine a 75cm vertical strut of 2in 10 SWG tube is fixed via a half inch hole in the top and a large self tapping screw below. This is used to brace the arms in the vertical plane, using tendons of 2mm Dacron to their tips and 3mm Dacron to an anchor near the centre of the spine. In the horizontal plane the arms are braced by 2mm Dacron between their tips and by the elements.

The spine is fixed to an 8mm plate 300mm x 150mm by a pair of U bolts. The plate is drilled for fixing to a rotator, replacing an upper clamp. It may otherwise be turned 90° for bolting to a rotating mast. Phasing lines terminate on polythene insulators attached each side of the plate. The lines then connect to a pair of pillar insulators, with one line transposed. The 450Ω rotation feeder also connects here. The other end connects to the main feeder at insulators on a cross arm 1m below the rotator. This is also where the upper mast guys are attached. A plastic bowl was fixed to the mast below the rotator to prevent the rotation feeder from touching but this was later replaced by a stand-off insulator.

The balun was designed for use with a linear amplifier when required to drive the open feeders of doublet antennas. Several friends have used this design with their G5RV antennas. It is a compact version of the linear 4:1 type used between HF broadcast transmitters and curtain antennas. Two 90cm lengths of PF100 coax, a pair of Amidon T200-2 cores and a weatherproof box with coax connector and two terminals are required for its construction. The cores are taped together and one length of coax is wound through them to give 7 turns. The other length of coax is coiled to fit into the box. **Fig 4.17** shows how the cables are connected. For clarity only one turn of the 180° cable is shown. The shell of the coax connector should be joined via a short thick conductor to a local ground point. The balun works as follows. Common mode signals appear equally on the 0 and 180° terminals. Signals on the 0° terminal are conveyed directly to the input connector. Signals on the 180° terminal are inverted to cause cancellation at the input terminal. The two transmission lines are in parallel on the input side and in series on the output side and therefore provide a 4:1 impedance ratio. The screen of the 180° coax forms an inductor in parallel with the input terminal and determines the lowest operating frequency. PF100 cable was chosen because its screen is solid copper foil and with the iron dust core it makes a low loss inductor. There is in theory no load impedance at which balance is not maintained and no upper frequency limit. Because no ferrite material is used the power handling capacity and working SWR are limited only by the current rating and insulation of the cables.

The Skeleton antenna has a wind load area of about 0.42 square metres, the same as a TH3, and G3LNP's has survived two winters with winds up to 145KMH. Its weight is less than 10kg but the total weight of rotator and mast is likely to be more than can be safely lifted without mechanical assistance.

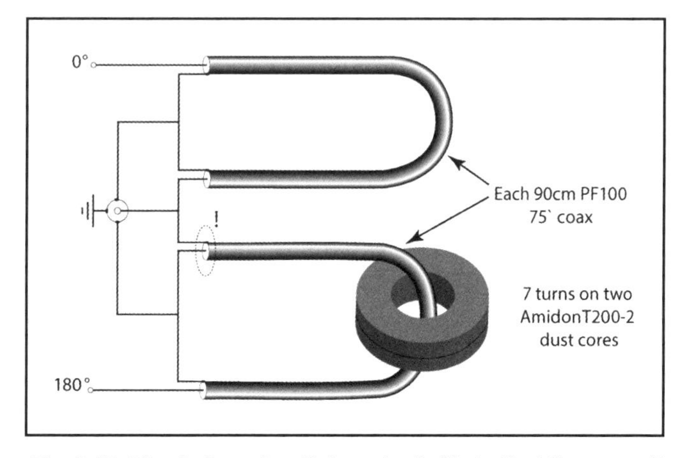

Fig 4.17: The balun circuit (see text). Note that there are 7 turns of the '180°' cable on the toroid, not one as shown here for clarity.

G3LNP used a scaffold pole derrick and 500lb winch. His Skeleton is fixed on a 12m scaffold mast where, compared with a half wave type multiband vertical antenna, it gets reports of 2 to 4 S-points in its favour.

If your tuner is unable to cope with the high SWR on the lower bands it may be reduced significantly by arranging to switch a capacitor across the phasing line junction. Typical values are 25pF at 14MHz and 500pF at 10MHz. In G3LNP's version these were omitted because he wanted to drive the antenna from a remote location and could not incorporate the extra switching facility. His Skeleton works on all bands, 10 to 29MHz, when using the internal automatic tuner of an Icom IC-765 transceiver and with an ITT / Macay MSR4020 automatic tuner, both without needing the capacitors. An Icom AT-180 tuner in the 'through inhibit' mode needed the capacitors and some fiddling with the length of the coax between balun and tuner before it would accept all bands. When using a linear amplifier he uses a manual 'T' network tuner.

A FOLDING YAGI FOR 144MHZ

This is a description of a standard Yagi: the novel bit is in how it is made foldable to allow for easy transportation for portable operations. It was described in the January 2013 *RadCom* by Eric Lake, 2W0WXM. He wrote, "Visiting a car boot sale one day I spotted an FM broadcast 4-element Yagi that had all its elements fixed in place with wing nuts. It was folded up and standing vertical, only taking up about 15cm square except for the driven element, which was a separate item. I wondered whether it would be possible to adapt this antenna for 2m, capitalising on its apparent folding capability. As the asking price was only £1 I bought it and, on returning home, started thinking."

Element	Distance from Reflector	Overall length
Reflector	0	517mm
Driven	310mm	see text
Director 1	480mm	470mm
Director 2	695mm	465mm

Table 4.6: 2m Yagi dimensions.

The basic construction of the antenna was a boom with three metal cross-pieces to which the refector and director elements were attached using wing nuts. A plastic insulator provided support for the folded driven element. All of the elements were made from aluminium that had been rolled into a tube shape, but not welded. The ends of the reflector and directors were flattened and rounded, presumably for safety. The elements could easily be shortened and the spacings adjusted to make the antenna work on 2m. Some research identified the dimensions in **Table 4.6**. The antenna was disassembled and suitable holes drilled in the boom for the elements at the new spacings. The reflector and director elements were shortened to the new lengths with new holes drilled for the wing nuts. **Fig 4.18** shows the general arrangement of the revised antenna elements and their spacings.

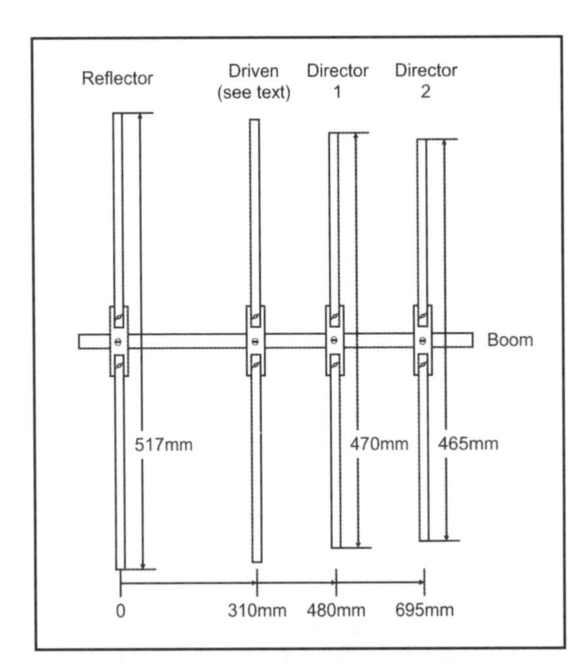

Fig 4.18: Main dimensions of the foldable antenna.

The key to making the whole antenna foldable was to work out a way of hinging the driven element. Given that, like the other elements, the two ends of the driven element were already attached using bolts and wing

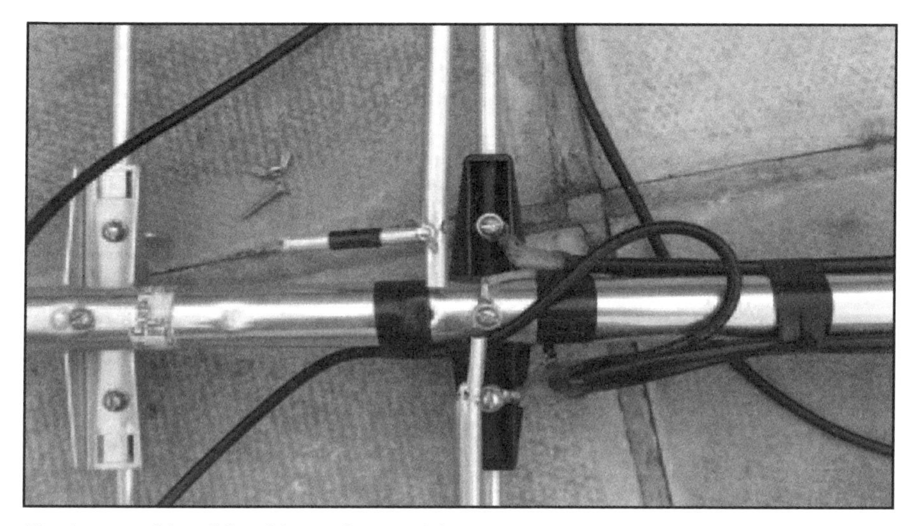

Final assembly of the driven element hinge.

nuts, it seemed sensible to work out a way of making the middle of the 'solid' side fold. This was done by cutting out a section of the tube and replacing it with a short section of 7mm solid aluminium bar. The size was chosen so it would slide inside the driven element tube. The fixing centres for the bar are the same as those for the 'open' ends of the driven element, 67mm. The photo shows the final assembly and **Fig 4.19** the dimensions. A file was used to round off the ends of the bar.

To allow for the hinge bar to fold at 90° it is necessary to cut out a section of the element tube. The bit that you cut out is shown in **Fig 4.20**. Put the tubing in a vice with the drilled hole upwards and hacksaw down about 5mm past the drilled hole. Discard the cut off piece then prise the slot open just enough to open up the tubing to allow the hinge bar to reach the 90° angle needed for it to fold.

The next thing to work on is the length of the driven element. Cut off the existing folds of the driven element and replace them with a section of solid aluminium bar folded into a U shape. The diameter of the bar was chosen so that it would slide inside the original tube, like a trombone, so that the length could be fine-tuned for lowest SWR. **Fig 4.20** shows the final dimensions and drilling details. It's important not to drill the U shaped piece until the antenna

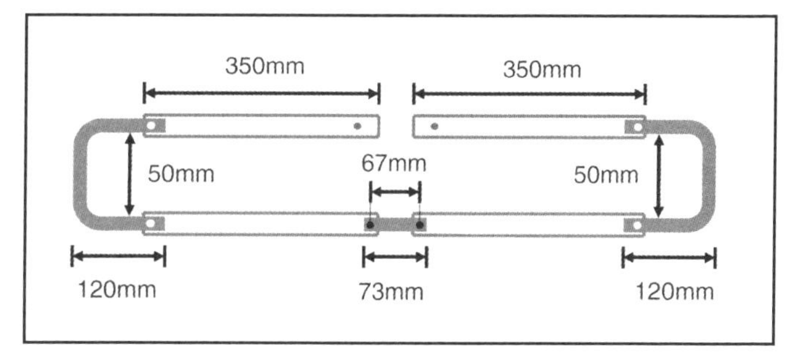

Fig 4.19: How the driven element was constructed. The thick solid lines represent aluminium bar, the parallel grey lines the original element material, suitably cut to length. The feed point is at the top.

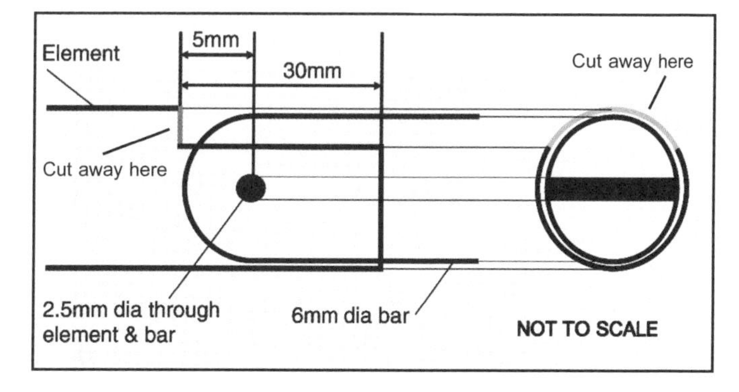

Fig 4.20: Final dimensions and drilling details of the driven element bar.

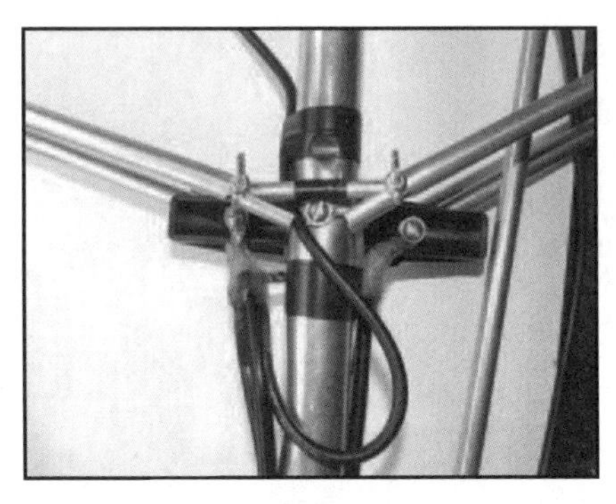

Another view of the driven element hinge. Note the insulating tape over the central bar.

Mounting clamp, made from two exhaust clamps.

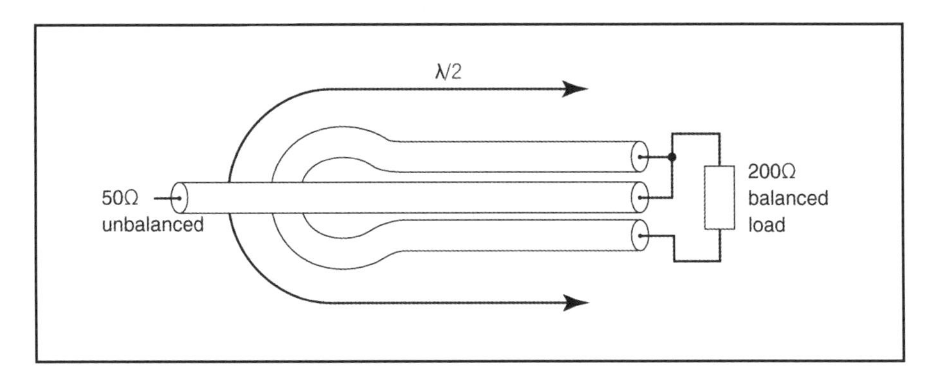

Fig 4.21: Balun details (see text).

is finished and has been adjusted for lowest SWR. The photo (above left) shows a close-up view of the finished assembly.

The balun used was a coaxial 4:1 type, see **Fig 4.21**. The loop should be an exact electrical half wave length (remember to allow for the velocity factor of the coax), using 50Ω or 75Ω, and preferably low loss, coax. 2E0WXM used 75Ω cable. When fixing the balun and feed line on the boom, run it symmetrically under the boom and to the rear. Attach to the boom with insulating tape and use terminal tags to connect the cables to the wing nut bolts of the driven element. The coaxial ends were made waterproof with silicone.

To mount the antenna on a mast, fit two exhaust clamps together with a short piece of tubing between them (see photo above right). The mounting arrangement is flexible enough to allow the antenna to be used with horizontal or vertical polarisation. It can also be tilted up or down according to requirements.

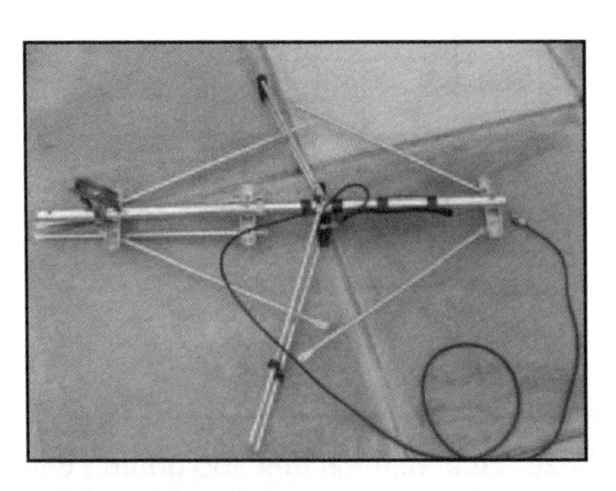

Folding the antenna.

Once the antenna is completed it can be tuned for lowest SWR by putting it on a short mast and adjusting the two U sections. Make sure that the two sides of the driven element are the same length. Once you have found the positions for lowest VSWR you can drill through the U bars and bolt them firmly to the driven element tube. The antenna built by 2E0WXM showed a 1:1 SWR across the 2m band.

The objective here was not to produce a beam antenna with superb performance, but to produce one that could be folded up in a very short space of time and be easily and conveniently carried. 2E0WXM wrote, "The idea does work and I'm pleased to have a beam antenna that can be folded into a size that easily fits into a single tent bag [see photo below – *Ed*]. It only takes a couple of minutes to open up and it's ready for use."

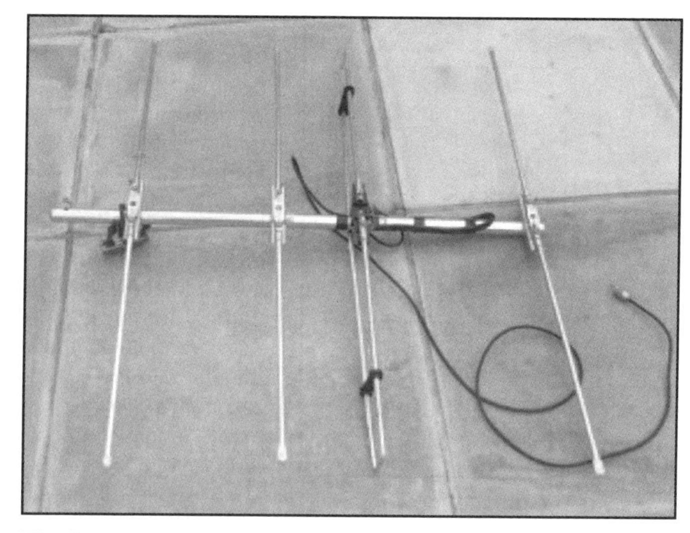

The foldable antenna fully open and ready for use.

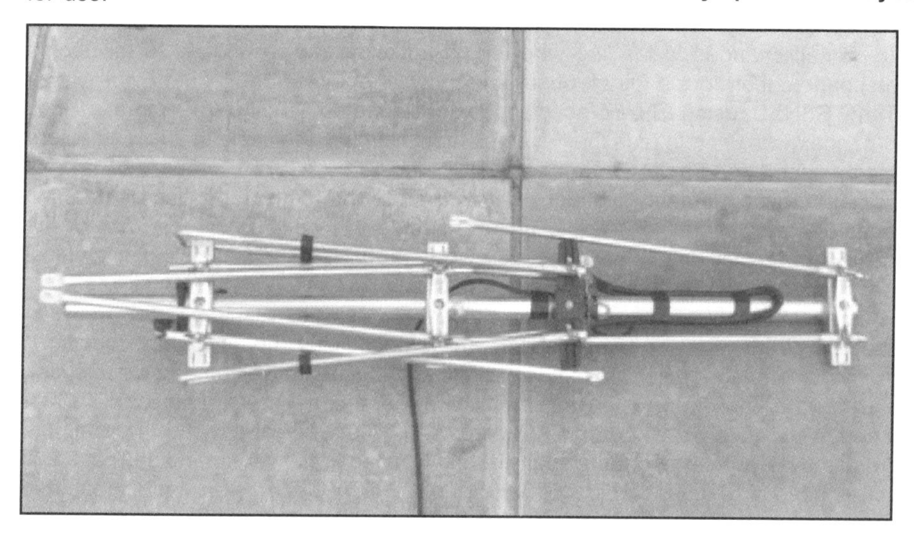

The antenna completely folded.

LINEAR LOADED QUAD

Linear loading is not a new technique, but it is mainly used on dipoles and Yagi-type beams (especially so on several of the commercially-made shortened 40m 2-element Yagis known as 'Shorty-Forties'). However, it is fair to say that, while not unknown, the technique is less used on loop antennas or quad beams. The late David Courtier-Dutton, G3FPQ, used linear loading on a 2-element 80m quad but even though the technique reduced its size somewhat, it must still have been a monster antenna and a sight to behold in the Surrey countryside!

For many people, though, even a 2-element quad for 20m can be too big an antenna to contemplate. This was the situation Bob Cox, G3PLP, found himself in. He wrote (in the March 2006 *RadCom*): "I am fortunate to have a multi-band quad from 17m to 6m, but have always been limited on 20m due to the proximity of a tree to the tower; there's not room to rotate it, and only regular pruning makes 17m possible." Doublets, delta loops and sloping dipoles were used on 20m to good

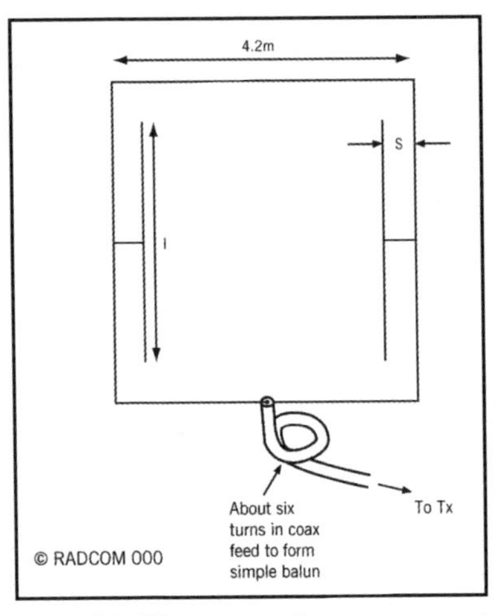

Fig 4.22: The general arrangement discussed in the text.

effect, but obviously did not provide the performance that could be obtained from a quad. With the approaching sunspot minimum [in 2005/6 – *Ed*] a better signal was required on 20m, so why not linear-load both the driven element and reflector of the 17m quad? The general arrangement is shown in **Fig 4.22**.

G3PLP first built a full-sized 18MHz loop from 16SWG copper wire and mounted it horizontally about 1m above ground. Each side was 4.2m (13ft 9in), and the measured feed impedance was about 120Ω. Several experiments were carried out by attaching loading wires to the voltage points using various lengths and spacings, running the loading wires parallel with the sides. Small spacings had little effect on loop resonance. However, with a spacing of about 15cm (6in) and a wire length of about 3 – 3.5m, it was possible to bring the loop resonance in the 20m band.

Another interesting side effect was that, as the amount of loading increased, the feed impedance at resonance reduced and a good match was obtained into 50Ω cable. Hence, a 20m loop in the same area as a 17m loop.

Small changes in loading wire length make sizeable changes to the loop's resonance, therefore a fair bit of cut-and-try was required. Subsequent analysis by Tony, G3NXC, using *EZNEC* showed the spacing to be equally, if not more, critical. 50Ω coax was used to feed the driven element, winding the cable in about a six-turn loop near the feed-point to form a simple balun.

The driven element resonant point can readily be measured using an antenna analyser, the amount of loading being adjusted to give a minimum SWR at the desired frequency (in G3PLP's case this was 14100kHz). The quad is a conventional design supported by glass fibre spreaders.

The loading wires are supported by tying them, using thin synthetic rope, between the end of the wire and the spreader, whilst maintaining the necessary spacing.

Now for the reflector. Adjustment of the loading length was made by measuring the front-to-back ratio using a remote dipole and a separate receiver. The first attempt produced a wonderful F/B ratio at 14600kHz: not exactly the optimum frequency. However, more loading wire and winding the tower up and down several times gave a resonance at 14100kHz and a F/B of better than 20dB measured using the remote dipole and an attenuator, between 14150 and 14200kHz. A reasonable F/B ratio was achieved over the whole band and the forward signal varied by little more than 1dB. The graph of **Fig 4.23** shows the SWR and F/B ratio variation across the band (G3PLP had no facility to measure forward gain). Interestingly, the reflector resonance for maximum F/B ratio was found to be about 3.8% lower, approximately 13500kHz.

Exact dimensions of the loading wires will have to be found by experiment as length spacing and diameter all seem quite critical, but the resulting antenna should prove quite effective for those who don't quite have room for a full sized 20m antenna.

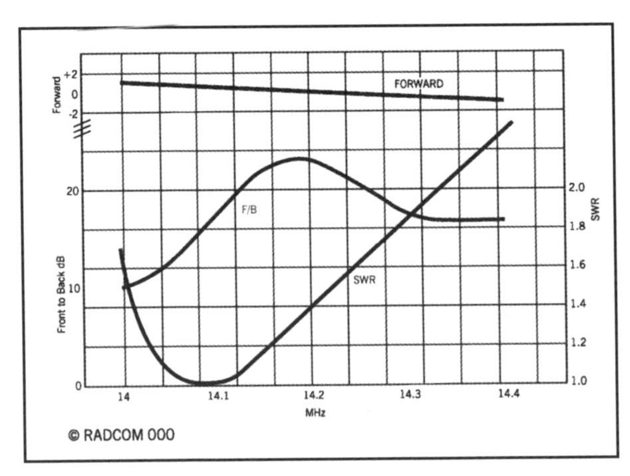

Fig 4.23: Variation of forward power, SWR and front-to-back ratio with frequency.

DOUBLE EXTENDED ZEPP YAGI

You've heard of the Yagi, you've heard of the Double Extended Zepp, but this novel designs combines the two into – well – a Double Extended Zepp Yagi. It was designed and built by William Alsup, N6XMW, and described by him in the October 2014 *QST*.

First, some background on the Double Extended Zepp. It is a centre-fed antenna consisting of two 0.62λ sections placed end-to-end (1.24λ total) and fed in the middle. The antenna has a pattern broadside to the wire with one major forward and one major backward lobe, each having about 3dBd gain at the cost of narrow beamwidths in the main lobes. That specific length maximizes the gain and has a feedpoint impedance that is easily transformed to 50Ω by a well-chosen segment of ladder line.

But, if you want only one direction, consider the Double Extended Zepp Yagi, which uses the Double Extended Zepp as the driven element and Yagi-type reflectors. After some trial and error *EZNEC* modelling showed that there should be *two* collinear standalone reflectors (rather than a single reflector) each a little longer than 0.5λ, connected in the centre by non-conductive cord as shown in **Fig 4.24**. All of the wires are 18AWG 'silky' wire (part number 532 from 'The Wire Man' (www.thewireman.com). The lighter-coloured lines indicate UV-resistant cord. Six insulators can be seen in appropriate locations.

Each of the antenna driven elements is an uncut continuation of homemade ladder line, which avoids the need to solder the ladder line to the antenna. Each starting wire length must be long enough to include one side of the radiating element as well as one side of the ladder line (plus a little extra for trimming in the tuning process). Ladder line is lightweight, which reduces sagging, and can be low loss.

The ladder line terminates 3ft above the ground in a 'Budwig' (www.budwig.com) antenna connector (see photo). The 5in gap between the two radiating sides at the top of the ladder line is maintained by thick monofilament, 'weed-eater' (garden strimmer) cord or PVC pipe.

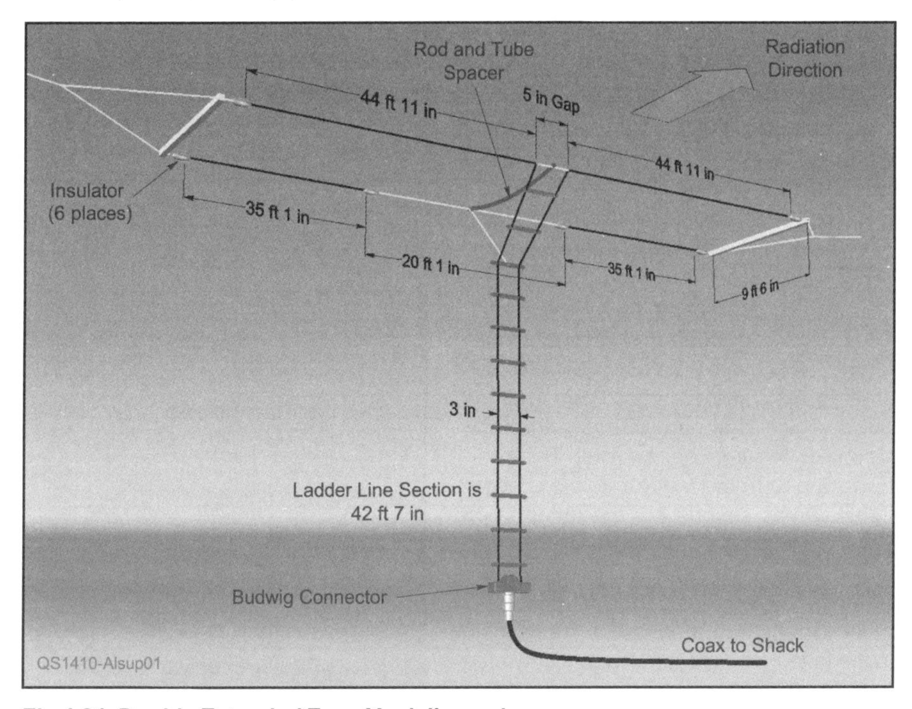

Fig 4.24: Double Extended Zepp Yagi dimensions.

Detail of the ladder line transition to coax and 'Budwig' connector.

The ladder line spacing is not critical. N6XMW used 3in spacing, using thin-walled irrigation tubing for spacers. The *length* of the ladder line, however, *is* critical. It needs to be a multiple of 0.5λ at the centre design frequency plus an additional matching length to compensate for the high impedance feedpoint of the antenna. The exact length of the ladder line needs to be adjusted by trial and error for best match.

As seen in **Fig 4.24**, each side of the driven element is 44ft 11in long, separated by a 5in gap, so the total Double Zepp length is 90ft 3in. Each side of the ladder line and each driven element are made from a continuous 87ft 6in length of 18AWG 'silky' wire (plus some extra for trimming). The reflectors are 35ft 1in each, with a cord in the 20ft 1in gap between elements. Spars maintain a 9ft 6in separation between the driven and reflecting elements. The 42ft 7in long ladder line terminates in a Budwig connector, then to 50Ω coax cable. You can insert a balun between the ladder line and the coax. N6XMW's antenna is 47ft above the ground.

The two spacers at each end of the antenna are lightweight wooden 10ft long, 1.5 x 1.5in spars. The middle spacer needs to be light to reduce sag, so two 6ft long, 0.25in diameter plastic rods were used, slipped (for stiffness) into a 6ft long aluminium 3/8in outside diameter tube for a total 12ft length. Taking into account the tendency to bow, this maintains the needed 9ft 6in spacing.

To hang the wires, run a line from the two tips of each wooden end spar, forming a triangle. At the apex connect a hoist line, and run it through a pulley suspended from a tree or mast. Instead of allowing the ladder line to hang directly down from the driven element, it should be drawn slightly towards the rear of the antenna, so as to hang down from the centre of the overall ensemble. Connect a drawstring from the ladder line about 12ft below the feedpoint to the centre spar at the two reflectors, as shown in **Fig 4.24**. This also helps to keep the antenna level.

EZNEC modelling shows a strong forward lobe with modest rearward and side lobes. The forward lobe gain is 6dBd, which is about 3dB more than the Double Extended Zepp by itself. The front-to-back and front-to-side ratios are about 12dB. On-air reports put the antenna on par with N6XMW's Cushcraft A4S Yagi on a 50ft tower. The Double Extended Zepp Yagi is inexpensive, performs reliably, is easy to maintain, and has suitable gain in the favoured direction. It does, however, take up a lot of space in the long direction, and needs two tall trees or masts.

Chapter 5
Receive Antennas

TWO-ELEMENT HORIZONTAL EWE

This antenna was designed to solve a major reception problem at the location of Floyd Koontz, WA2WVL, in Florida and was described by him in the December 2006 *QST*. His desire was to maximise the received signal-to-noise ratio (SNR) of European 80m SSB signals (around 3.8MHz).

WA2WVL had previously designed the vertical EWE antenna and described it in articles in *QST* in February 1995 and January 1996 (the antenna gets its name from its resemblance to an inverted U). However, he discovered that vertically-polarised antennas did not work well at his Florida location and decided to try the EWE in the horizontal plane. He discovered that it worked very well.

With vertical EWEs, a single element has a deep null off the back. A single horizontal EWE, on the other hand, would have that null at 0° elevation instead. Over real ground, the single horizontal EWE has a front-to-back ratio (F/B) at 3.8MHz as shown in **Fig 5.1** for 30° elevation. The low angle gain increases with height above ground, as with most horizontal antennas, but the pattern remains nearly the same at 10 to 30 foot (3 – 9m) heights. The size can be chosen to fit the available space but four supports (trees, towers, house, etc) are needed to hold up the corners.

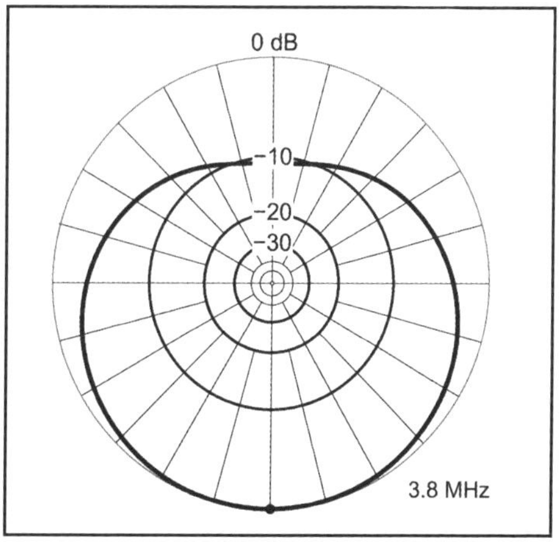

0 dB
-10
-20
-30
3.8 MHz

Fig 5.1: *EZNEC* modelled pattern of a *single* element horizontal EWE at 3.8MHz and 30° elevation.

Of the different sizes and shapes that can be used, one is optimum. It is a square, λ/8 on a side (a total of λ/2 around). For this size (30 x 30ft / 9 x 9m for 3.8MHz), the feed is non-reactive and can be matched with a simple broadband transformer. The calculated feed impedance was 1337Ω and was matched with a 26:5 two-winding transformer. The F/B was more than 11dB on 80m SSB and 15dB on 160m. The front-to-side ratio was only about 4dB.

A two-element design was therefore established to improve the back and side rejection. Available trees allowed a spacing of 100ft (30m). **Fig 5.2** shows the layout of this array including the location of the feedpoints and terminations.

Fig 5.3 is the modelled pattern of the two-element array, again at 3.8MHz and 30° elevation. As is evident, the two-element array resulted in a sharpened beamwidth and a significantly improved F/B. While the pattern for 30° elevation is shown, it is similar with lower output at lower elevations. The antenna was to be 25ft (9.6m) off the ground. This was determined to be the maximum height that I could safely reach from my 24ft (7.3m) extension ladder. Vinyl insulated 14 gauge 'speaker wire' was used, but any wire strong enough to be pulled tight without stretching could be used. The loops were tied at the corners (every 30ft / 9m) by 1/8in (3.2mm) Dacron ropes.

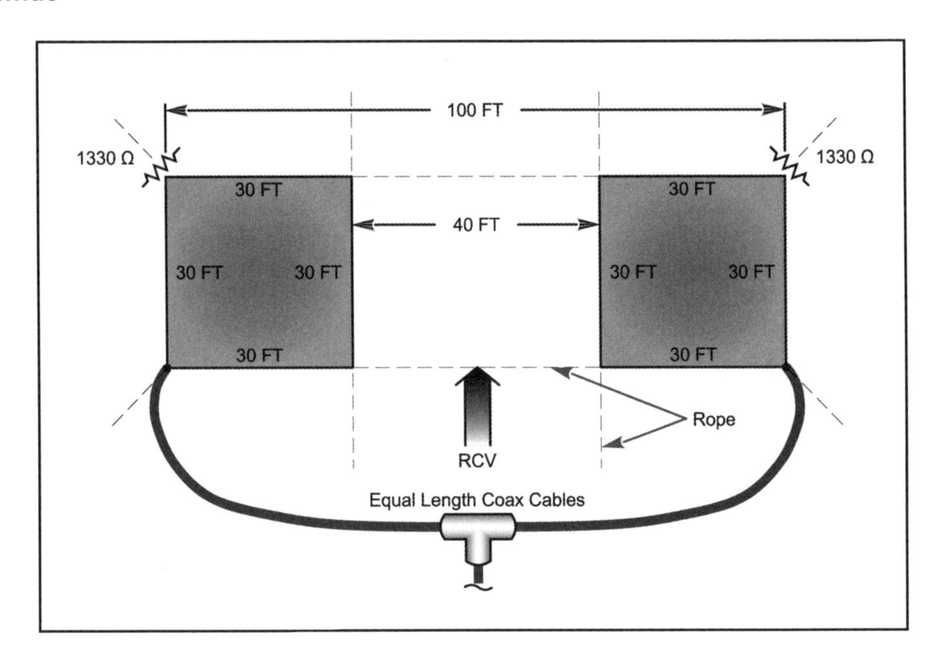

Fig 5.2: Orientation of the two-element horizontal EWE antenna array.

The two loops must be fed in phase to develop the pattern and there is a 50% chance of getting it right. If you guess wrong, the pattern will have a 30dB null off the front and receiving performance will be poor. Both feed points had equal length coax cables attached (I used 100ft / 30m for each run length). A BNC T-connector was used to parallel the 100ft coax cables and the third port connected to the coax that went to the shack.

The calculated gain of the two element horizontal EWE (at 20º) at 3.8MHz was about -18dBi (a $\lambda/4$ vertical has a gain of about 0dBi). For a height of only 10ft (3m) the gain drops to -24dBi. Most modern receivers and transceivers have sufficient sensitivity so that even at this level there is sufficient gain that the signal-to-noise ratio is limited by external noise. If the noise level doesn't increase by at least 10dB when replacing a dummy load with the receive antenna, you will need to have additional gain to get full benefit from this antenna. In order to compare directly the performance of this antenna with your transmitting antenna, you may want to add gain so that the signal levels are similar and the S-meter moves in the same way. A preamplifier with a gain of 18 to 25dB should be just what you want for 80m, more if you use the antenna on 160m. It might be that switching your receiver's preamp on provides just the amount of gain required to bring the signal level up to that from your transmitting antenna, and that is a suitable way to equalise the gain.

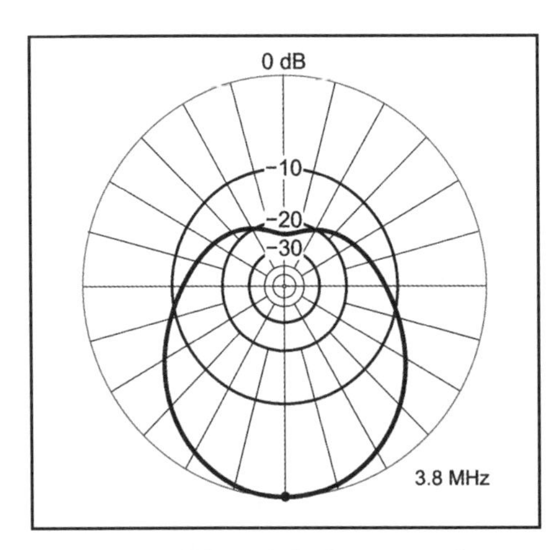

Fig 5.3: *EZNEC* modelled pattern of a two-element horizontal EWE configured as in Fig 5.2 at 3.8MHz and 30° elevation.

Operation with this system has been gratifying with the SNR often being improved by 10 to 20dB on 80m compared with my reference antenna. On 160m, many SSB stations can now be copied that were below the noise before.

ORTHOGONALLY STEERED (OS) ANTENNA

Directional receiving antennas, such as small loops that can be steered in azimuth, have traditionally been used to help separate radio signals that arrive from different directions. However, if two transmissions share both frequency and bearing, this technique cannot be effective. If the signals originate from different distances relative to the receiver and are propagated via ionospheric reflection, however, they will generally arrive at different angles of elevation. An antenna having a reception pattern that is steerable in the *vertical* plane can then help to separate the signals. In the general situation of receiving in the presence of interference, if signals arrive from different distances, but on the same channel, we need an *orthogonally steered*, or OS antenna.

Such an antenna was designed and built by Tony Preedy, G3LNP, as a receive antenna for the bands from 160 to 30 metres inclusive and described by him in the October 2010 *QST*. G3LNP used the K9AY loop, a terminated triangular wire loop antenna that can improve reception on the lower HF bands by virtue of its directivity, as the basis of his design. [The K9AY loop was originally described in the September 1997 *QST*: 'The K9AY Terminated Loop – A Compact, Directional Receiving Antenna', by Gary Breed, K9AY – *Ed*].

Fig 5.4 shows a typical azimuth plot for a K9AY loop at 45° elevation with a termination of 470Ω. The K9AY loop is popular because it is quieter than many antennas designed primarily for transmitting on 160m and the lower HF bands. To cover more than one band usually requires careful choice of both antenna dimensions and the termination impedance. Commercial versions are available that use pairs of reversible crossed loops to give four null directions. The effectiveness of such antennas is dependent on both the vertical and horizontal arrival angles of the interference.

A computer simulation of the K9AY antenna indicated that the optimum termination impedance depends on frequency, ground conductivity and resistance of the ground connection. In one commercial version this dependence is accommodated by making the termination resistance variable from the operating position. By making the termination a complex impedance (reactance plus resistance), not only can the frequency range be extended and compensation made for ground conditions, but the null can be controlled in the elevation plane as shown in **Fig 5.5** and **Fig 5.6**. Here we see null elevations of 20° and 60°, respectively. The corresponding horizontal patterns remain substantially the same as that shown in **Fig 5.4**. Once you have control of the termination impedance, the

The OS antenna control box sits underneath the Yaesu TV rotator.

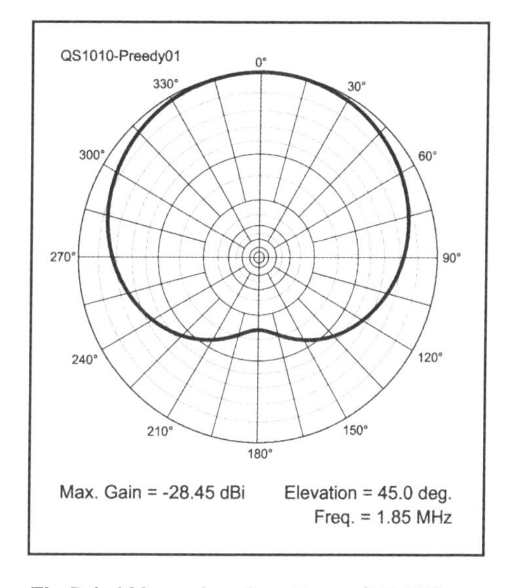

Fig 5.4: 160m azimuth pattern of K9AY loop at 45° elevation.

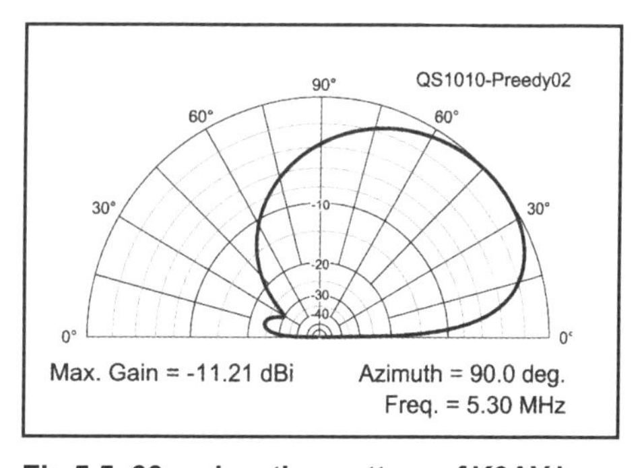

Fig 5.5: 60m elevation pattern of K9AY loop terminated for maximum front-to-back ratio at 30° elevation.

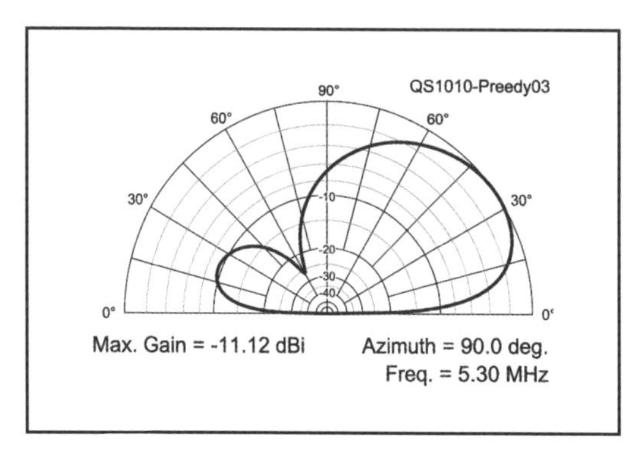

Fig 5.6: 60m elevation pattern of K9AY loop with termination adjusted for elevation null in rearward lobe.

dimensions of the antenna also become less critical. G3LNP's modelled loop had an apex height of 20ft and base of 30ft. In this case, the null could be steered from 0° to at least 80° elevation over the range 1.8 to 10.2MHz. Above 85° an unrealisable negative resistance is required. A null depth greater than 60dB could be achieved for a wide range of ground parameters and for anticipated values of ground connection resistance. **Fig 5.7** is a plot showing the typical relationship between termination impedance R +*j*X and null elevation of the modelled antenna.

By replacing the usual termination resistor with a wideband transformer and transmission line, such as is used on the feed side, it is possible to provide a remote variable termination in which values are modified both by the square of the turns ratio and by the transformation due to the length of the transmission line. A wide range of reactance values down to 0Ω can be obtained from a series resonant L-C circuit as it is tuned away from resonance. This principle is used in the remote termination unit of the OS antenna. If combined with a rotation system it is thus possible to place a null on an interfering signal arriving from any likely angle of azimuth or elevation.

The OS antenna does not require a large area or much height. A version of half the size of my design will be suitable for 3.5MHz and above if you accept 12dB less sensitivity. **Fig 5.8** shows the circuit of the termination unit and how it interfaces to the loop. The unit is built in a metal screening box that fits below the rotator controller. The X control is a 470pF air dielectric variable capacitor. The inductor, nominally 30µH, for 1.8MHz, consists of 32 turns of 0.5 mm (24AWG) enamelled copper wire, close wound on a 32mm diameter photographic 35mm plastic film can. Taps are provided at 23, 19, 16 and 13 turns, with unused turns shorted by a single pole

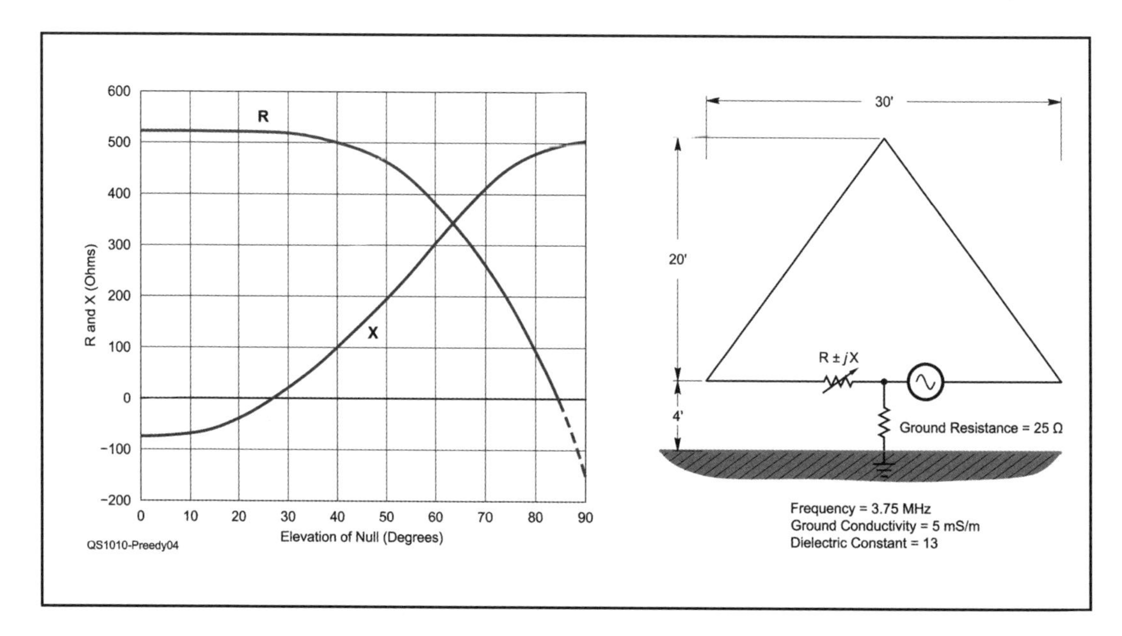

Fig 5.7: Plot of angle of elevation null versus complex termination impedance at 3.75MHz.

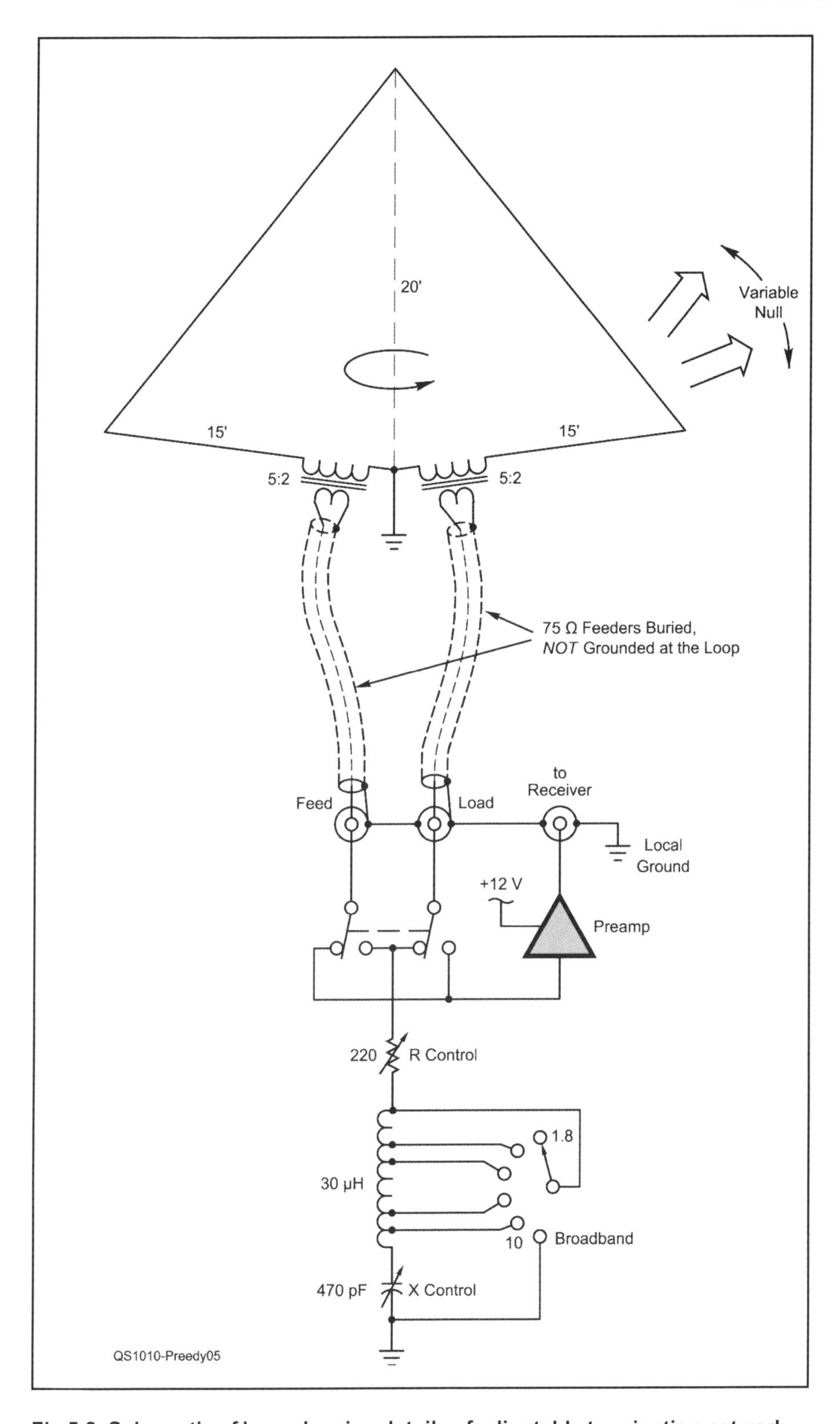

Fig 5.8: Schematic of loop showing details of adjustable termination network.

Antenna base mechanical construction details showing TV rotator for azimuth control.

switch, for the 80, 60, 40 and 30m bands, respectively. A sixth switch position shorts the inductor and capacitor for less critical, resistance only, wideband operation. A 220Ω non-inductive linear variable resistor serves as the R control. A double pole reversing switch is used to swap feeders.

The antenna was made rotatable by attaching it to a Yaesu TV antenna rotator, as shown in the photo. This rotator has filters built into the servo system that help make it immune to transmitted fields. The use of an AC motor avoids commutator interference. The plastic netting and bricks were added to prevent rabbits from chewing on the cables. The rotator is fixed to a 4ft length of steel pipe driven vertically into the soil with a 6in projection. This has its lower end flattened to resist rotation but is not connected to the copper ground system because of the risk of electrolytic corrosion.

Four 4ft ground rods at the corners of a 3ft square copper sheet are used for the loop's ground connection. This is all that is necessary because a better ground connection only makes the feed VSWR excessive for higher elevation null angles.

The horizontal arms of the antenna are each made from aluminium tubing salvaged from an HF Yagi antenna. The dimensions tapered from 7/8in to 1/4in. These and the 20ft vertical pole are clamped to a central 8in square plate of 5/8in thick SRBP (Paxolin) insulating material using pairs of U-bolts. Almost any rigid insulating material, even varnished plywood, will be suitable here. The U-bolts set the slope of the arms to 10°, to put their tips about 3ft above ground. G3LNP used another piece of the Yagi, extended with a 4ft insulating bamboo top section, for the vertical pole. The pole, which is not an active part of the antenna, is insulated from the rotator casing. The SRBP panel is fixed to the upper rotator clamp with 6mm bolts by using the holes intended for the rotator's upper U-bolt. The sloping arms use 45ft of wire: the wire diameter is not electrically significant.

The antenna transformers are wound with 5 turn primaries and 2 turn secondaries on twin hole ferrite cores, about 5/8in square, housed in a small plastic box fixed to the SRBP panel. Cores should be the smallest obtainable, wound with thin insulated wire to minimise static capacitance between windings. G3LNP measured his at 5pF but commented that even this is probably too much for optimum performance on the higher frequencies.

A significant potential problem with this type of relatively insensitive antenna is the effect of pick-up on the outer conductor of the feeders. This can introduce unwanted signal voltages into the radio via the ground connection impedance or the capacitance between transformer windings. Cables should therefore be buried and the outer conductors not grounded at the antenna but where they emerge at or near the radio. The rotator cable was decoupled by winding as many turns of cable as possible through a 2in ferrite toroid near the rotator before it joined the buried coaxial cables. Because the motor only consumes 10W, the cable is thin six-core alarm type that allowed 16 turns on the ferrite core.

Feeders are 75Ω foam dielectric TV type and should not exceed 200ft in length or losses may make it impossible to get sufficient range of load control on the higher frequencies. Waterproof PF100 type cable is recommended if they are buried directly. G3LNP's are in an MDPE water pipe for protection from moles. Identical feeders allow switched interchange between load and feed connections so that the antenna can be reversed without rotation or load adjustment.

Operation

Be sure to connect the OS control unit in a receive-only path to the radio! If your transmitting antenna is nearby it may be necessary to ground the feeder to protect the 220Ω resistor and pre-amplifier while transmitting. You may also need to detune the transmitting antenna when receiving in order to achieve the deepest null. Relays controlled by the PTT line can automate these functions if desired.

If R is set to 1/3 of maximum, around 75Ω, and the band switch is at the BROADBAND position, the antenna termination will be near 500Ω, regardless of feeder length or frequency. These settings should allow the antenna to work as originally designed by K9AY. Use this condition to find the bearing of a signal you want to eliminate, or for general listening. As the controls are moved from these settings the load will become complex but actual values will be influenced by feeder length unless it is a multiple of 180° at the operating frequency.

Adjustment is straightforward. Just select the band, rotate the loop for minimum interference, or use the great circle bearing if you know the direction of the interfering signal. Now adjust the R and X controls successively to further minimise an unwanted signal or noise. Use the reversing switch to satisfy yourself that the antenna is effective – or to confirm that the interference is still there!

Forward gain with 180ft of feeder, assuming a 2:1 VSWR, was predicted to vary from –28.6dBi at 1.8MHz to –6.4dBi at 10.1MHz and this appears to have been achieved. On all bands noise from the OS antenna exceeded that due to the receiver. Additional amplification above that from the preamplifier in the radio may not have been necessary if the antenna was only used in the forward direction. In this application, while receiving off the back, an additional amplification stage may be needed, especially on the lower bands.

Results

The antenna was very effective at locating the source and reducing the effect of local interference. Because of the intermittent nature of amateur signals it was found that AM broadcast signals were more useful while getting familiar with the controls. The antenna worked well in the broadcast band with the switch in the 160m position. Daytime ground wave signals at MF could be reduced by more than 60dB, such that it was possible to completely separate a distant radio station from a closer one sharing the same frequency. The reversing switch then gave a choice of programmes, both free of interference. Daytime rejection was often more than 50dB on 1.8MHz, falling to about 30dB at 10MHz. In late afternoon S9 German signals on 1.8MHz could be reduced to inaudibility while being able to copy more distant Polish stations on the same frequency and bearing. This was also tried on 3.7MHz with a pair of French stations, one in the north and one in the south, both on the same bearing. Again, the vertical null could be controlled to attenuate one relative to the other. On 3.7MHz the daytime sky-wave signals from stations in Jersey, Scotland, Ireland and the Isle of Man could all be reduced by more than 40dB without significant loss on closer near vertical incidence skywave (NVIS) signals.

As dusk descended, results on 1.8MHz remained good. East European stations could still be rejected, but it was found that 7MHz propagation at that time was unstable, making it difficult to obtain a consistent null because the apparent arrival direction was changing. On a 7MHz East European broadcast station, for example, the front-to-back ratio would change rapidly from a worst case of 10dB to troughs of 40dB, even though in the forward direction fading was only a few dB.

After midnight, when 7MHz propagation was more stable, a 30 to 40dB rejection of Russian and other East European stations was obtained. Before dawn, listening on the transmitting antenna to an eastern USA net on 1.85MHz, it was obliterated by a Russian station on that frequency. Using the OS, it was possible to choose which contact to listen to.

Near sunrise, on 3.7MHz, it was impossible to reduce an Italian by more than 20dB, presumably because there was more than one propagation mode, whereas at the same time USA stations could be reduced by about 30dB.

With experience and logging of termination settings I could estimate arrival elevation of signals and follow this as propagation changed. The loop was not quite as quiet as a Beverage antenna – but this loop can be rotated.

The total cost, using UK sourced materials, was about US $400, which seems reasonable for a 1.8 to 10MHz rotary beam. This included the rotator, hardware, pre-amp, cables and all components except the parts salvaged from the Yagi. Probably surplus CB antennas, bamboo or fibreglass poles with added conductors could be substituted.

THE DK6ED DOUBLE LOOP ARRAYS

A receive-only antenna is a must for DXing on the lower frequency bands. Transmit antennas are very large for these bands and they pick up lots of noise, so low-level signals are lost. To improve signal-to-noise ratio (SNR) a receive antenna should have a narrow horizontal and vertical beamwidth as well as a good front-to-back (F/B) ratio.

The Beverage is an ideal receive antenna – but only if you have enough space! Not only should it be *at least* one wavelength long (so, for 1810kHz, 165 metres or 545 feet in length) it also has to be orientated in the correct direction, as signals are received off the end of the wire.

The rotatable DK6ED double loop array.

Magnetic loops – sensitive to the magnetic field – are very small compared with a wavelength, and are particularly useful in noisy areas. They receive equally well from both edges of the loop, but have a null on each side of the loop. Terminated loops such as the flag, pennant, EWE, and K9AY loop are directional (a design for a two-element horizontal EWE is featured at the beginning of this chapter). Chris Kunze, DK6ED, has come up with some novel designs which he described in the March 2015 *QST*. He combined pairs of terminated (magnetic) loops, and phased them so that the radiation pattern has enhanced directivity for improved SNR and a good F/B ratio. One of the designs is rotatable, and the other is a fixed array.

Each of the double loop designs is a compromise between mechanical

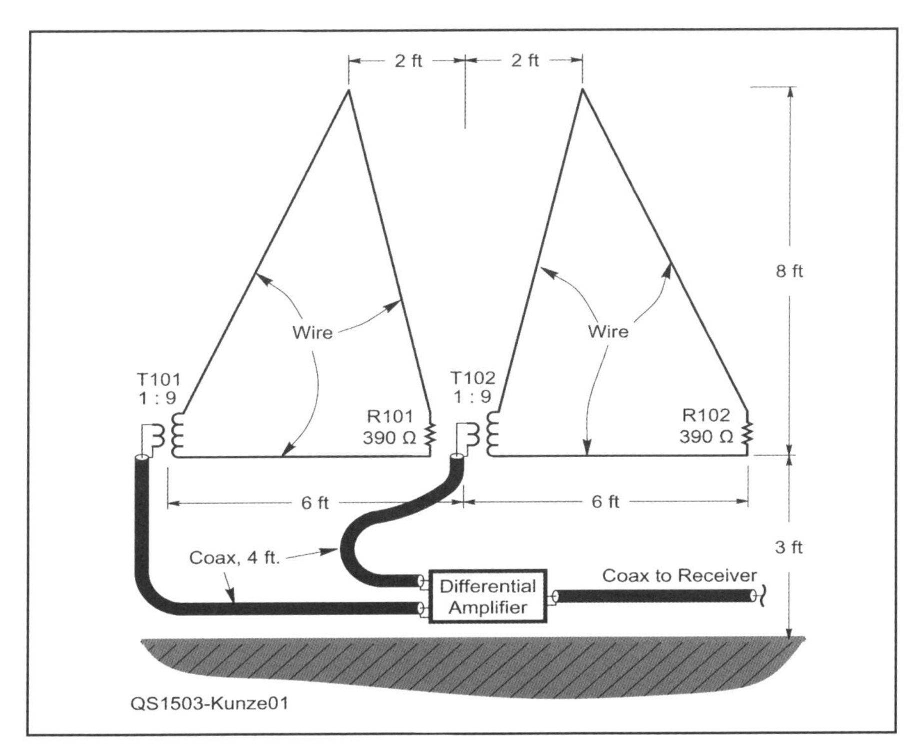

Fig 5.9: Rotatable double loop wiring diagram. R101, R102 – 390 W resistor; T101, T102 – primary 2 turns, secondary 6 turns on BN73-202 ferrite core (www.kitsandparts.com); Preamplifier – see text.

dimensions and an optimised radiation pattern. The rotatable system is shown in the photo and its wiring diagram in **Fig 5.9**.

The other antenna is used in a fixed position. The radiation pattern of the fixed-position double loop system is similar to the pattern of a one-wavelength-long

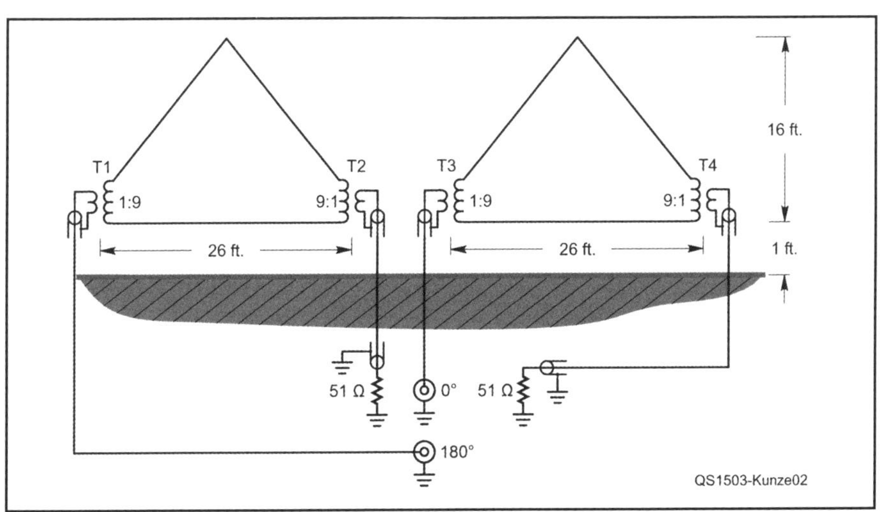

Fig 5.10: Double loop receiving antenna for fixed use. R201, R202 – 82W resistor; T201 – T204 primary 2 turns, secondary 6 turns on BN73-202 ferrite core (www. kitsandparts.com)

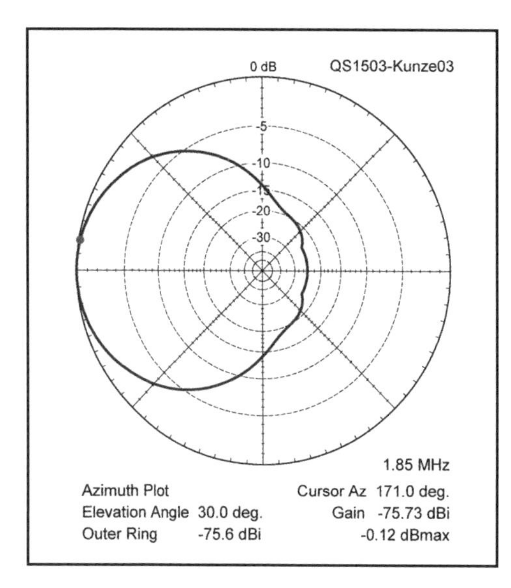

Fig 5.11: Azimuth pattern of the rotatable double loop receiving antenna.

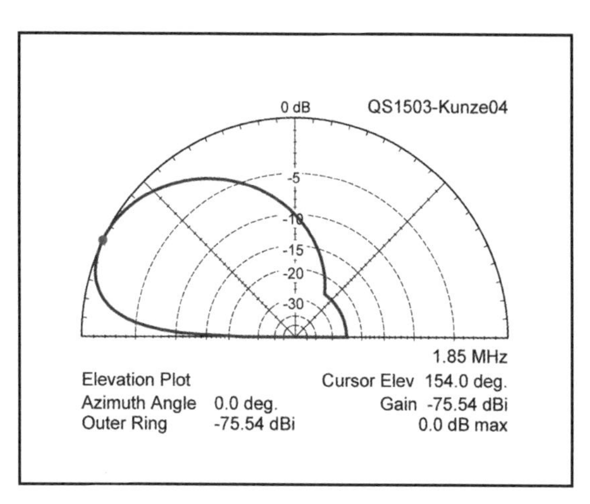

Fig 5.12: Elevation pattern of the rotatable double loop receiving antenna.

Beverage antenna, but needs only 52ft of space (**Fig 5.10**), compared with the Beverage length of 515ft.

Because of their small size, the signal output of the DK6ED system is low, so you will probably need to use a preamplifier. The SNR of the double loop system is higher than what you get from a single terminated loop, and the SNR does not depend on the output level. The double loop signals are combined in a differential preamplifier, or with a 180° hybrid.

Using *EZNEC*, DK6ED found the smallest antenna dimensions where the F/B ratio began to deteriorate below his chosen minimum limit of 24dB. The resulting antenna is quite small, so can be turned using a small TV rotator. The antenna is built on a simple wooden frame. The two transformers and the two resistors are housed in small plastic boxes at the ends and centre of the lower support. Take care to build the two loops in a loop pair symmetrically. Make sure the feed lines have equal lengths and that the transformers are connected in the same winding sense.

Despite the small size of this antenna, the azimuth (**Fig 5.11**) and elevation (**Fig 5.12**) patterns have narrow beamwidths. As mentioned above, the signal output is quite low compared with conventional receive loops, so a preamplifier is used.

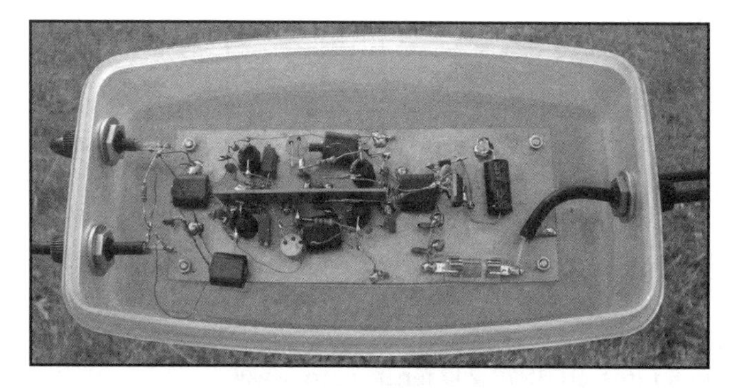

The preamplifier mounted in a plastic box.

Since these antennas are broadband, for the RF Preamplifier DK6ED opted for an amplifier designed by Dallas Lankford and available on the website of J Miles, KE5FX (www.thegleam.com/ke5fx/norton/lankford.pdf)

DK6ED used a differential configuration that provides the necessary 0° and 180° inputs as shown in **Fig 5.13**. The gain of the Norton stage is about 15dB. The preamplifier was mounted in a box (see photo) at the base of the

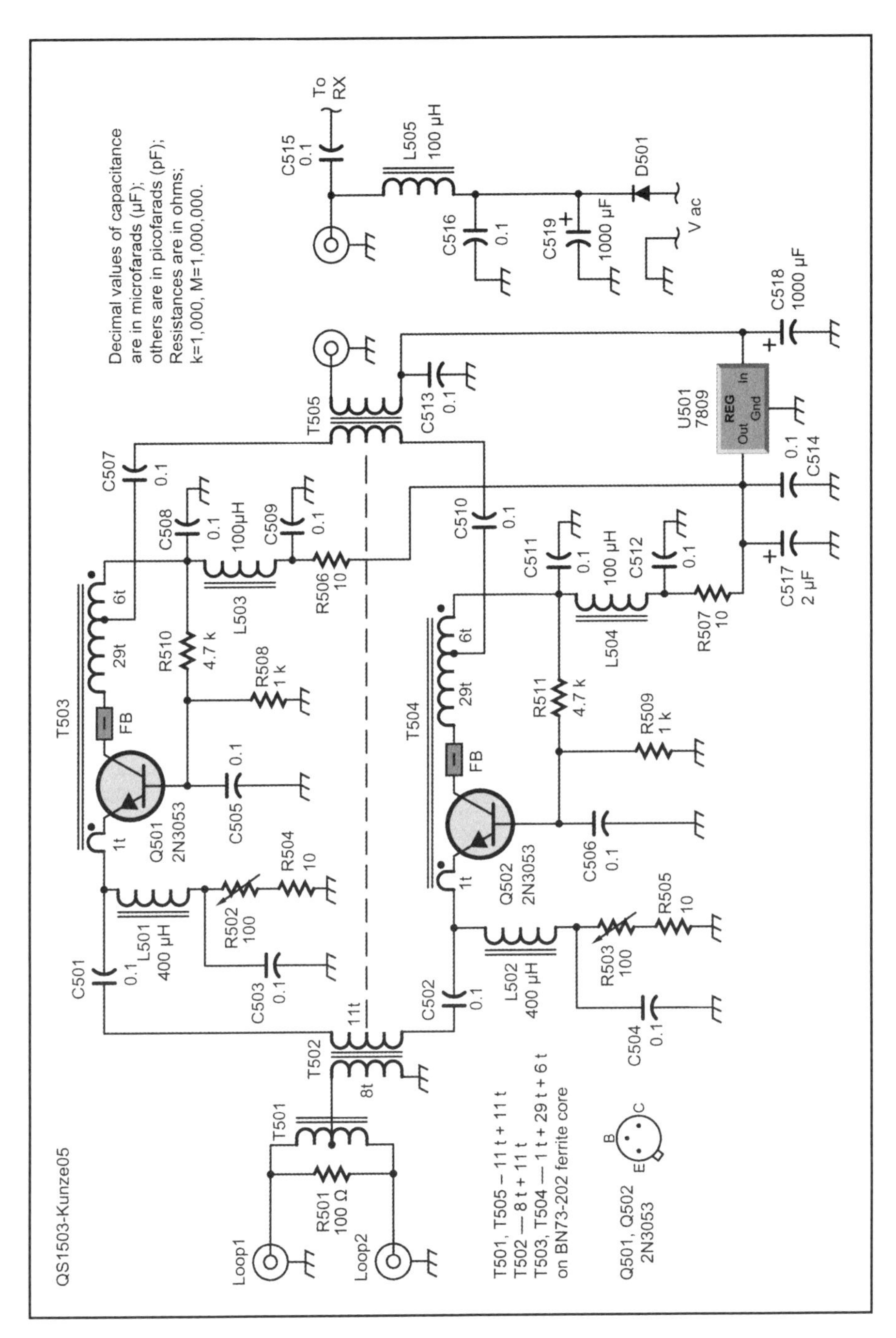

Fig 5.13: Preamplifier schematic. C501 – C516 0.1µF capacitor; C517 2µF capacitor; C518, C519 1000µF capacitor; D501 silicon diode; L501, L502 400µH inductor; L503 – L505 100µH inductor; Q501, Q502 2N3053 transistor; R501 100Ω resistor; R502, R503 100Ω variable resistor; R504 – R507 10Ω resistor; R508, R509 1kΩ resistor; R510, R511 4.7kΩ resistor; T501, T505 11 turns + 11 turns on BN73-202 ferrite core (www.kitsandparts.com); T502 3 turns + 11 turns on BN73-202 ferrite core; T503, T504 1 turn + 29 turns + 6 turns on BN73-202 ferrite core; U501 7809 regulator; FB Ferrite Beads.

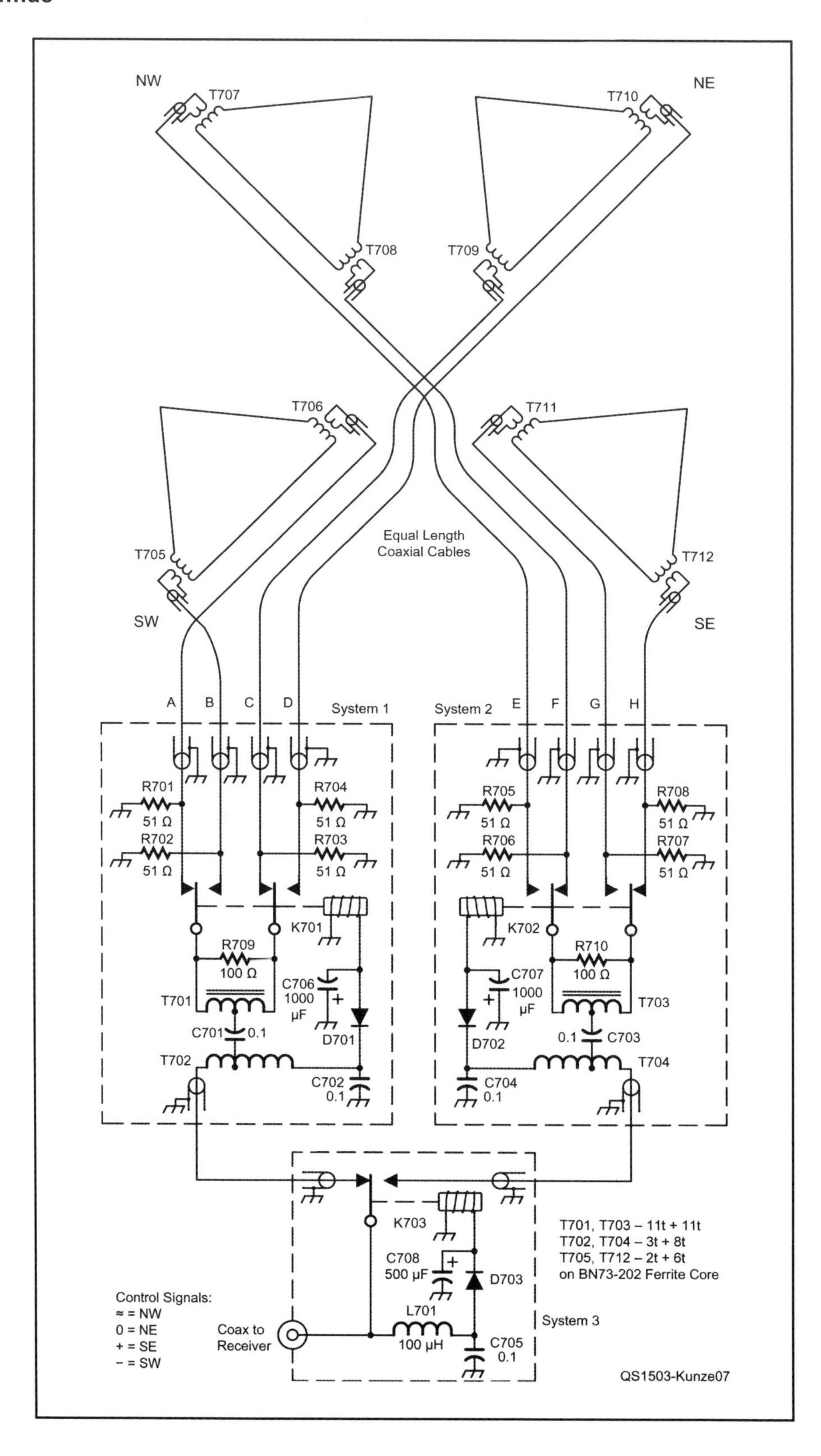
NW
T707
NE
T710
T708
T709
T706
T711
Equal Length
Coaxial Cables
T705
T712
SW
SE
A
B
C
D
System 1
System 2
E
F
G
H
R701
51 Ω
R702
51 Ω
R704
51 Ω
R703
51 Ω
R705
51 Ω
R706
51 Ω
R708
51 Ω
R707
51 Ω
K701
K702
R709
100 Ω
R710
100 Ω
C706
1000
μF
C707
1000
μF
T701
T703
C701
0.1
0.1
C703
D701
D702
T702
T704
C702
0.1
C704
0.1
K703
C708
500 μF
D703
T701, T703 – 11t + 11t
T702, T704 – 3t + 8t
T705, T712 – 2t + 6t
on BN73-202 Ferrite Core
Control Signals:
≈ = NW
0 = NE
+ = SE
– = SW
Coax to
Receiver
L701
100 μH
C705
0.1
System 3
QS1503-Kunze07

Fig 5.14: A pair of bi-direction loop pairs covers four principal directions. C701 – C705 0.1μF capacitor; C706, C707 1000μF capacitor; C708 500μF capacitor; D701 – D703 silicon diode; K701, K702 DPDT relay; K703 SPDT relay; L701 100μH inductor; R701 – R708 82Ω resistor; R709, R710 100Ω resistor; T701, T703 11 turns + 11 turns on BN73-202 ferrite core (www.kitsandparts.com); T702, T704 3 turns + 8 turns on BN73-202 ferrite core; T705 – T712 2 turns + 6 turns on BN73-202 ferrite core.

---

antenna and is powered through the coax feed line from the shack. Be careful with the phasing of the feedback coils in the Q501 and Q502 circuit as the amplifier might oscillate. You can check it with an RF probe. The current at each section should not exceed 5mA. Otherwise, simply twist the loop around to change the phase of the feedback. For alignment, just equalise the currents through R504 and R506 while connected to a 9V source.

Common mode currents on the feedline can be suppressed by winding the coax over a large toroid (such as an Amidon FT 240-77) close to the receiver.

The small rotatable system is useful for point-to-point operation, or when looking for a station in a known direction. However, for random contacts the fixed system can switch directions instantly. The double loop shown in **Fig 5.10** is the basis of a fixed-position array. In his article 'The K9AY Terminated Loop – A Compact, Directional Receiving Antenna' (*QST*, September 1997) Gary Breed, K9AY, showed how an array of two single loops can cover NW-SE and NE-SW. DK6ED's array uses two identical double loop pairs and control boxes, one each for the NW-SE and the NE-SW directions. Each control box contains the relays and a combiner. The two control boxes are linked in a third box (**Fig 5.14**). Build the loops symmetrically, and pay attention to the sense of the transformer windings. All of the feedlines should have equal length. The antenna patterns in each of the four principal directions are in **Fig 5.15** and **Fig 5.16**.

DK6ED summarised his project by writing, "My fixed array system performs best because of its narrower beamwidth and better F/B ratio, and I can scan rapidly

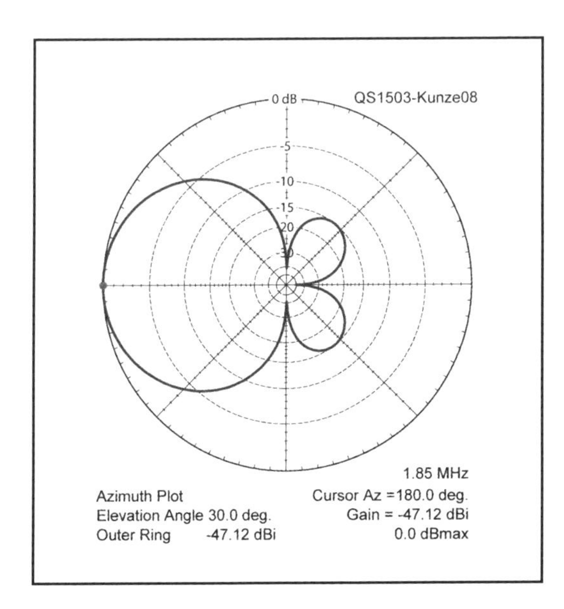

Fig 5.15: Azimuth pattern of the fixed double loop receiving antenna.

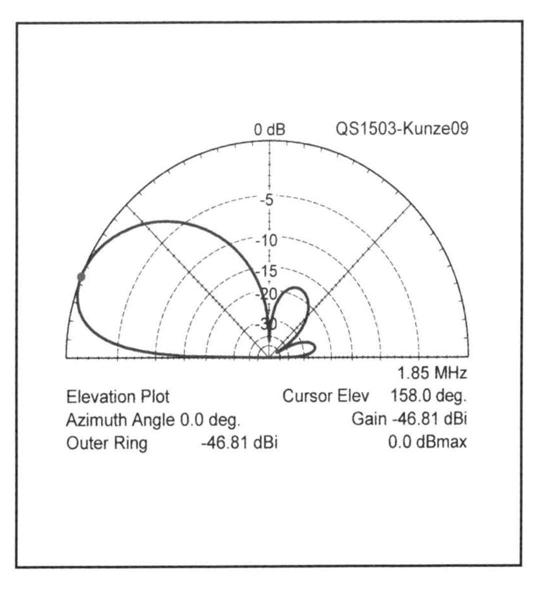

Fig 5.16: Elevation pattern of the fixed double loop receiving antenna.

in azimuth by just switching the loops. The rotatable system is very easy to build and can be carried around looking for the location with the lowest local noise level. In actual use, my new double loops improve the SNR by as much as 25 dB. I experience F/B ratios of 20 – 30dB, as predicted from the simulations.

"Any kind of double loop in a noisy area will, of course, not out-perform a Beverage in a low-noise area, but nevertheless I enjoy significantly improved reception."

Chapter 6
Other Novel Antennas

THIS CHAPTER IS made up of those antennas that do not fit comfortably into any of the other chapters – they're not really dipoles, verticals or loops. It includes a few 'controversial' antennas, such as the Cross Field Antenna (CFA) and the E-H antenna. We don't want to get into a discussion about *how* an antenna works, but in a book about 'novel' antennas the CFA and E-H antenna could hardly be totally excluded. Not all the antennas in this chapter have the dubious distinction of being labelled 'controversial', though. We start with a novel mobile antenna that was apparently originally designed for use on board ships.

THE ROOF RACK MOBILE ANTENNA

Like many other mobile operators, Peter Dodd, G3LDO, having used a car roof rack as a support for HF and VHF antennas over the years, speculated whether the *roof rack itself* could be used as an antenna. He concluded that the most likely candidate for the basis of an antenna design based on the roof rack was the Directional Discontinuity Ring Radiator (DDRR) antenna.

In G3LDO's 'Antennas' column in the May 2007 *RadCom*, Peter quoted Tom Francis, NM1Q, who had written a short history of this antenna in *QST*: "The DDRR was designed by J M Boyer for shipboard operation at very low frequencies. Basically, the DDRR is a quarter wave, end-fed antenna, grounded at one end and shaped as a single-turn coil. (Mr Boyer described his antenna design in the January 1963 issue of *Electronics* as a 'hula-hoop' antenna.) The DDRR requires a counterpoise (ground plane) for effective operation. The antenna was described to amateur radio operators in the June 1970 issue of *73 Magazine* by W E English, W6WYQ. This article illustrated the problems confronted in building an HF version of the DDRR and how these problems were overcome. After numerous requests for a ground-mounted version, English wrote an article for the December 1971 issue of *QST*, in which he described a 40m DDRR as 'an apartment dweller's dream'." The DDRR was included in the earlier editions of *The ARRL Antenna Book* (see **Fig 6.1**), but has been omitted from later editions.

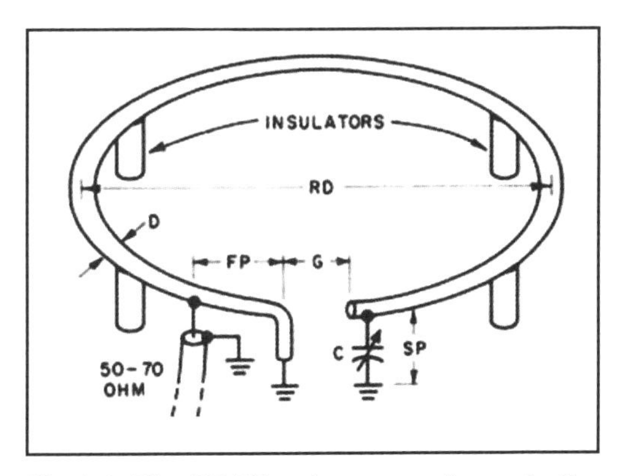

Fig 6.1: The DDRR antenna as shown in the 14th edition of *The ARRL Antenna Handbook*.

These early articles attracted the interest of W2WAM, who wrote a mathematical analysis of the DDRR antenna ('A study of the DDRR Antenna', Robert B Dome, W2WAM, *QST* July 1972), the conclusion of which was that increasing the vertical height could improve the efficiency appreciably.

G3LDO considered the DDRR just to be a short vertical, with the rest of the quarter wavelength formed into a horizontal loop. He built a mobile version of this antenna from 2mm copper tubing to look exactly like a car roof rack. However, instead

The G3LDO Mk 1 experimental Roof Rack Mobile Antenna. The far end of the element is supported by a plastic block, which could be slid along the element to alter its height above the roof of the car.

of using a circular horizontal section his design used a rectangular configuration. There were two reasons for this approach: firstly, a square or rectangle can be made up using straight sections of tubing, with the ends joined using 90° angle joints. Secondly, a rectangular configuration looks more like a roof rack, and if constructed well enough can even be used as a roof rack (although not as a transmitting antenna at the same time!) The Mk 1 mobile DDRR antenna is illustrated in the photo above.

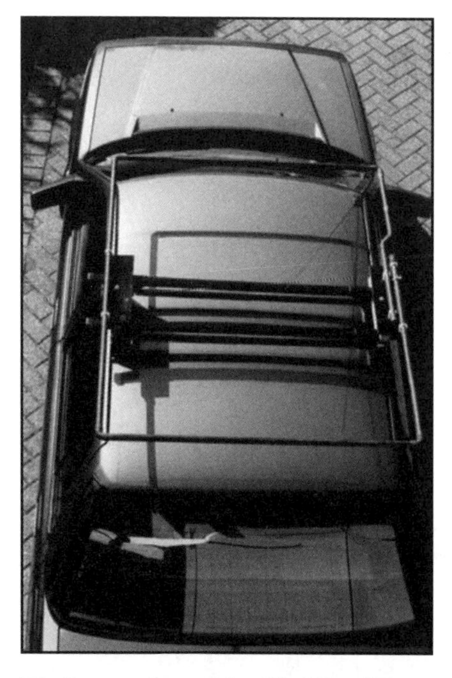

Bird's eye view of the Mk 2 Roof Rack antenna. The element overlap provides capacity for tuning the antenna rather than capacity to ground.

A more robust version was constructed and fixed to the roof of the car by a bar roof rack – the sort used to transport ladders or timber – and is illustrated in the photo, left. After reading W2WAM's mathematical analysis G3LDO increased the vertical height to 250mm (10in), which was as high as practicable if the antenna was to masquerade as a roofrack. The antenna is shunt fed by tapping the feedline centre up from the ground end of the element as shown in the photo opposite.

The traditional DDRR is tuned with capacitance to ground as shown in **Fig X.1**. G3LDO's variation uses capacitance provided by overlapping the top end of the element with the lowest part of the horizontal section. The capacitance is adjusted by sliding the top end of the element along the tube-to-wall fittings, thereby adjusting the overlap. These fittings offer just enough friction to retain the element in the desired position. Calibration marks are used so that the overlap position can be set at any frequency within the tuneable range.

This antenna proved to be a very useful single band antenna for 14MHz. It did not need to be removed when parked for fear that it might get vandalised or stolen, after all it looked just like a luggage rack. Furthermore it was possible to drive into low- roofed multi-storey car parks without having to remove it.

In an end-fed vertical the current in the

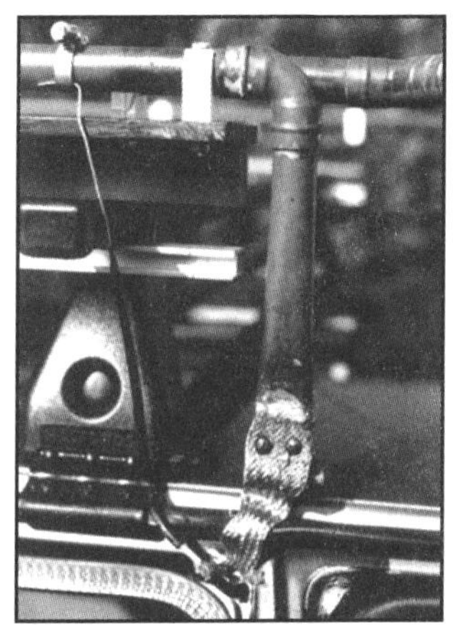

The feed end of the mobile DDRR antenna. The lower end of the element is connected to the bodywork of the car with tinned copper braiding, together with the braiding of the feeder coaxial cable. The centre of the coaxial cable is connected with a wire to an adjustable point (shunt feed) on the antenna element.

counterpoise is equal to the current in the antenna. This mobile version of the DDRR is no exception and in this case the counterpoise is the vehicle metal body.

RF current in the antenna and counterpoise is indirectly proportional to the radiation resistance. The DDRR antenna has a very low radiation resistance so the current induced into the vehicle body and vehicle wiring can be very high when using a standard 100W transceiver. It is this RF current in the vehicle body that can affect the running of modern cars that use microprocessor controlled engine management systems (see the notes in the sidebar for testing amateur radio equipment in a modern vehicle).

MOBILE RADIO & THE DDRR

Mobile operation is an excellent way of experimenting with compact and unusual antennas and the mobile DDRR antenna is a good example. But it is essential to ensure that your experiments do not compromise the operation of the vehicle. Here are some typical notes on things to check.

Initial Checks

Switch on transceiver and check that it and the antenna functions correctly (low SWR). Switch on ignition (but not engine) and check that all instruments, warning light displays are normal. Now transmit and verify that displays are unaffected, then repeat the test in all operating bands. If the transceiver is multimode, repeat the test with all modes. In each case use the maximum RF power. If there is *any* disturbance of the vehicle instrumentation, ***stop*** and identify the source of the problem before continuing. Repositioning the antenna and power leads is the sort of action that might solve the problem. If the above tests have been completed successfully proceed to *Static Checks.*

Static Checks

Start the engine of the vehicle and repeat all the tests described in the *Initial Checks*. Check that there is no disturbance of the engine control or engine speed. Switch on the vehicle lights, indicators etc while transmitting. Check that no unintended flashing or indication occurs. If the above tests are completed successfully proceed to *Mobile Checks.*

Mobile Checks

Find a road free of traffic. Start the vehicle and while moving slowly operate the transmitter. Check that brakes, etc all operate as normal. Repeat using bands, modes, etc as applicable to your transceiver. If all is OK, increase to normal driving speed and repeat the tests. If there is any unexpected reaction from the vehicle (accelerator, transmission, steering or other in-car electronic device) ***stop*** transmitting immediately. If these tests are satisfactory perform a braking test at normal speed while transmitting.

THE E-H ANTENNA

The E-H (or just EH) antenna is one of those 'controversial' antennas I alluded to in the introduction to this chapter.

Much of the development work of the E-H antenna was carried out by Ted Hart, W5QJR, but his website about the E-H antenna appeared to be down when this book was being researched and compiled. However, a 36-page e-book written by Ted Hart still appears on the Internet on a Russian website at ehant.qrz.ru/book.pdf The book offers an explanation by W5QJR of how the antenna works. He claims that "The EH Antenna concept is the most significant change in antenna theory in more than 120 years" and rebuffs arguments by those who say it cannot work.

Those interested in reading W5QJR's claims for this antenna should read the e-book but, for the opposing view, there is also a website by Charles Rauch, W8JI, which explains why he thinks the E-H antenna does not and *can* not work. This website is at: www.w8ji.com/e-h_ antennas.htm

Without getting into any argument about whether or not the antenna works as W5QJR claims, the sidebar explains the claimed characteristics of the E-H antenna and **Fig 6.2** shows the basic layout.

The E-H antenna can be home made from information provided in W5QJR's e-book. Steve Nichols, G0KYA, has built two – one for 20m and the other for 15m. In his book *Stealth Antennas* (RSGB, 2nd edition, 2014) Steve wrote, "They are very small, not very easy to set up, but they do work. The 20m one is about 4ft long, the 15m version is about 3ft long. I have made contacts on both bands with the antennas and at times they

CHARACTERISTICS OF E-H ANTENNA
(as claimed by W5QJR)

Size of an EH Antenna is typically less than 2% of a wavelength.

EH Antennas are typically vertical dipoles – no land required for radials.

Radiation Resistance is a constant 120Ω.

Efficiency of the EH Antenna approaches 100%.

For receiving, it produces the same signal level as a full-size Hertz antenna.

Signal to noise ratio is significantly better than a Hertz antenna.

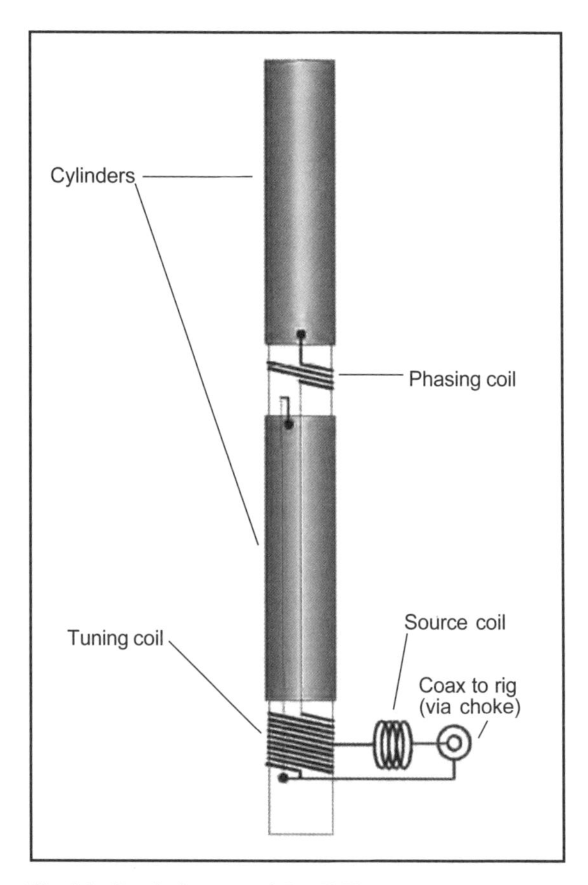

Fig 6.2: Basic layout of the E-H antenna.

have been equal to other more conventional antennas and at times they have been 2 – 3 S-points worse."

The commercially-made versions of the E-H antenna are manufactured by Arno Elettronica in Italy and information on them can be found (in Italian) on their website at www.arnoelettronica.com/menu/venus.htm

The Arno Elettronica E-H antennas have been independently tested in *RadCom* twice, first by the late Bob Henly, G3IHR ('The Arno Elettronica E-H Antennas', H R Henly, CEng FIEE MBCS, G3IHR, *RadCom*, September 2003). G3IHR tested the 20m and 40m versions of the E-H antenna and his conclusion (in part) was that "Both of the E-H antennas performed extremely well as general-purpose antennas, exhibiting no significantly different performance to my normal antennas [a trap vertical and a G5RV – *Ed*]... Both antennas worked quite well at ground level but only compared favourably with the other antennas when operated under similar conditions, i.e. at a similar height and position with relation to surrounding objects."

G3IHR then obtained the 80m and 160m versions of the E-H antenna for testing but sadly became a Silent Key before he could write up his results. The 80 and 160m antennas were later passed to Don Field, G3XTT, for his analysis. What follows is an abbreviated version of his review, which was published in the August 2005 *RadCom*.

Don wrote: "When I was asked if I would review the E-H antennas for 160 and 80m I approached the matter with great interest, but also some trepidation. The theory behind the working of E-H antennas is controversial, to say the least, and has been the subject of some quite intense correspondence in 'Technical Topics' and elsewhere. I have no intention of entering the technical debate, as better qualified people than me have already crossed swords on the subject. However, as an LF enthusiast [note that G3XTT is using the term 'LF' informally to refer to the 160 and 80m bands, not the low-frequency part of the spectrum or the 136kHz amateur band – *Ed*] I was keen to see whether, for those without the space for a full-size LF antenna, the E-H antenna would be a suitable substitute."

Both the 80 and 160m antennas are just 8ft in length, which is a far cry from the 67ft or so height required for an 80m quarter-wave vertical or 135ft length for an 80m dipole (and twice these dimensions for 160m).

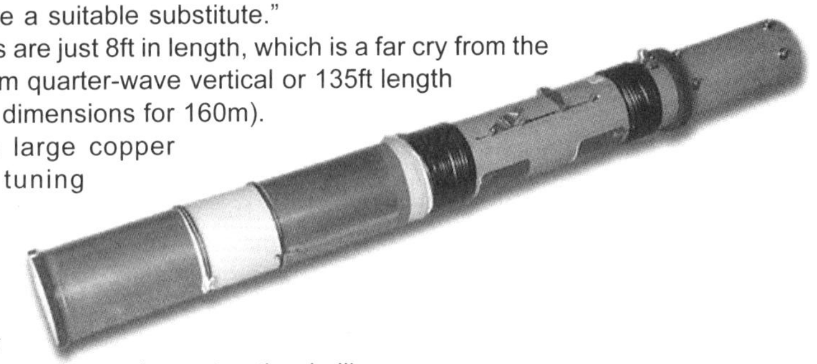

Imagine the antenna as two large copper plates along with various tuning components. To make the system more practical, the plates are rolled into the form of a tube, and the whole antenna encased in weatherproof plastic. The cut-away photo gives you some idea of what the internal construction is like.

Each antenna comes with brackets to fix it to a mast of up to 1.5in diameter, and all you then need to do is to connect a coax feed line (there is an SO239 connector at the base of the antenna). It is recommended that the antenna is well in the clear, avoiding metallic guy wires and other metal objects if at all possible. Personally, I would have preferred the LF versions to come with brackets that would fix to a 2in mast, as most of my hardware is geared around that size, and the weight is non-trivial (4.9kg for the 80m version and 5.5kg for the 160m version).

I should point out at this stage that each of the E-H antennas is single-band, so that, if you wanted to cover all nine bands from 160 to 10m, you would need nine antennas and nine, preferably widely-spaced, supports. In practice, many amateurs buy E-H antennas for the odd band(s) they cannot otherwise cover.

There is a single model for 80m, with a claimed 2:1 SWR bandwidth of 170kHz. An adjustable copper tuning sleeve around the antenna allows the tuning range to be adjusted to any part of the band. In my case I set the lower end of the 2:1 range

The 80m E-H antenna on the reviewer's tower, luffed over so that the antenna's tuning sleeve can be adjusted.

Size comparison: although tiny in terms of wavelength, at 8ft long the 160m and 80m E-H antennas are still fairly substantial objects and weigh in at 4.9 and 5.5kg respectively.

at 3500kHz, and found the 2:1 bandwidth to be 180kHz, i.e. slightly better than specification. SWR at resonance was close to 1:1. Power rating of both the 160 and 80m antennas is 2kW on SSB and CW and 500 watts on continuous modes (RTTY, AM). This should be more than adequate for use in the UK, and I found that I could run 400 watts into the 80m antenna without difficulty. To my surprise, I was able to adjust the tuning sleeve with the tower luffed over, and the resonance didn't change measurably once the tower was vertical and the antenna raised to about 35ft. I felt that a clear location at 35ft or thereabouts would be fairly representative of those with limited garden space. If you could get an antenna much higher you could start seriously considering a full-size inverted-L or something similar.

The specification claims that performance is within 3dB of maximum over a 350kHz range which means, theoretically, that you could cover the whole 80m band by adjusting the resonance to the centre of the band and using an ATU, but personally I would be reluctant to do this as there are likely to be very high voltages present at the antenna away from resonance and damage may ensue (see 160m comments, below).

I used the antenna on 80m CW, comparing its performance on both transmit and receive with my full-size 80m inverted-Vee dipole (centre at about 45ft). I was easily able to work around Europe, with good signal reports, putting some semi-rare DX into the log (ZB2FK in Gibraltar and ZA1AA in Albania, for example). Received reports suggested a one- to two-S unit difference between the E-H antenna and the dipole, and a more accurate test with a local amateur, using a calibrated attenuator, indicated that the E-H antenna was 6 – 8dB down on the dipole. I consider this a good result for an antenna that is so much smaller. My usual 'rule of thumb' is that performance starts to fall off rapidly once an antenna falls much below two-thirds of full-size, but the E-H antenna gave the lie to this. In comparison, I have been singularly unimpressed with small loop antennas that I have had the opportunity to try out in the past.

The 160m E-H antenna mounted on the reviewer's tower at 35ft AGL.

One of the claims for the E-H antenna is that it can often be better than a full-size antenna on receive, as noise pick-up can be lower. I didn't notice this effect, one example being V51AS (Namibia) who I worked on my dipole with solid if weak copy of his signal, but who was barely audible on the E-H.

My overall impression was that the 80m E-H antenna performed remarkably well for its modest size.

The 160m antenna is exactly the same size as the 80m E-H, and has a claimed 2:1 SWR bandwidth of 40kHz, and a 3dB bandwidth of 70kHz. Two versions are available, to cover the 1830 – 1850kHz range (some adjustment of the centre frequency is possible) or the high end of the band (nominally 1913 – 1933kHz). I used the 1830 – 1850kHz version, installed in exactly the same location as the 80m antenna had been, and extended my 80m dipole for 160m for comparison purposes. The resulting dipole had to be somewhat bent at the ends to fit in my garden, but this is probably a good comparison as that is the best that many users could manage on topband.

My initial measurements showed a 2:1 SWR bandwidth of just 20kHz (consistent with the 1830 – 1850kHz working range claimed in the brochure, though not with the claimed 40kHz 2:1 SWR bandwidth), with a best SWR of 1.6:1.

Initial results were extremely promising. Tests with my local amateur friend suggested that the E-H antenna was no more than about half an S-point down on the dipole. The two antennas were fairly close together (getting two topband antennas several wavelengths apart would require considerably more real estate than I have!) so there may have been a degree of mutual coupling. To overcome this, during the tests I removed the 160m dipole and replaced it with the 80m dipole and this had no discernible effect on the signal received from the 160m E-H. So as far as I could tell, any mutual coupling was insufficient to affect the overall conclusions.

The first few contacts made around the UK and Europe indicated that the E-H antenna worked well, with little or no observable difference between it and the dipole. I was also pleasantly surprised to notice (contrary to my experience with the 80m version) that received signals were much clearer due to less noise pick-up. Indeed, the first contact I made, with an OE7 station (Austria) would not have been possible on the dipole because he was inaudible under local noise.

Where I did have some difficulties was in experiencing some sort of flash-over at higher power levels, despite my having taken care not to exceed 400 watts into the antenna. In the brochure it is rated at 2kW for CW and SSB operation (though even the manufacturer had recommended keeping output power below 500 watts, so the brochure figure appears overoptimistic).

In conclusion, as a receive antenna the E-H allowed me to hear stations that were inaudible on my main antenna. Used at the 100 watt level, it would appear to give results remarkably close to a compromise 'full size' antenna and therefore probably no more than an S-point or two down on even a pretty good 160m antenna. This is far better than I had anticipated when I first saw the antenna and realised just how compact it was. However, I would be somewhat

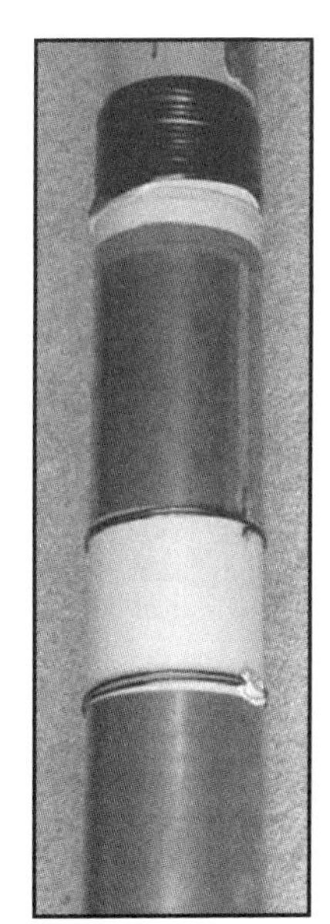

Another look inside one of the E-H antennas.

concerned at the relatively narrow bandwidth of the antenna and would be reluctant to use one at much more than the 100 watt level, for fear of damage.

Testing antennas can never be a truly objective process, unless one has access to a professional antenna range. The nearest I can come to this is by asking a local amateur, close enough not to be significantly affected by propagation effects etc, to undertake careful signal strength measurements. Beyond this it is a case of listening and making QSOs on the band, and trying to gauge how effective an antenna is, based in my case on some 37 years of LF operation. Unfortunately, during the period of the tests, low-band propagation and activity was disappointing, but I feel that I can make some reasonably informed comments. I came away from the tests favourably impressed with the E-H antennas, whatever the pros and cons of the theoretical debates that have been raging.

On that subject, the late Pat Hawker, G3VA (who made no secret of being sceptical about claims that antennas such as the E-H worked by Poynting Vector Synthesis), quoted a letter from Peter Martinez, G3PLX, in his 'Technical Topics' column (*RadCom*, December 2003). G3PLX wrote that the E-H antenna "is no more than a short fat loaded dipole matched to 50Ω". G3VA, however, said that "Personally, I would suggest that the E-H antenna is a novel form of the established 'Normal-Mode Helical Antenna' (NMHA) that can provide a reasonably effective antenna with a length of little more than 0.1λ. I recall the (in)famous 'Joy-stick' antenna of the 1960s, comprising a coil of wire wound on a broomstick..."

So while the debate on *how* the E-H antenna might work rages on, respected amateurs such as G3PLX and G3VA seemed to agree that there is no reason why it shouldn't work – but just not in the way claimed by its designers. While both G3IHR and G3XTT found that the E-H antennas they tested performed about as well as standard single-element vertical or wire antennas at modest heights, it is also only fair to point out that the commercial E-H antennas receive a very low rating on the www.eham.net/reviews website. Here, the majority of users found that their E-H antenna did not work, while a minority reported that it worked about as well as a dipole.

CROSSED FIELD ANTENNA (CFA)

It would be unfair to feature the E-H antenna in a book about novel antennas without also including the Crossed Field Antenna (CFA) which, in many ways, is the forefather of the E-H. The CFA was developed by Maurice Hately, GM3HAT, and his Egyptian student, Fathi Kabbary, at The Robert Gordon University in Aberdeen in 1985. They developed a theory called Poynting Vector Synthesis ('PVS') in which they proposed that transmitter power could be divided into two, one part to create the E field and the other the H field. They argued that the CFA they developed 'synthesises' the E and H fields in a time / phase relationship in order to transmit the radio wave. They claimed a greater efficiency than a conventional quarter-wave vertical in a structure *less than* 5% of a wavelength high (i.e. λ/20).

An MF broadcast CFA, showing the conical funnel and the 'E' plate (Wikipedia).

The CFA they developed consisted of three parts: a horizontal metal disc they called the 'D' plate, a vertical cylinder called the 'E' plate and, on top of these, a funnel-shaped metal lattice structure.

A number of these CFAs were installed in Egypt and elsewhere for use by medium wave broadcast stations and as far as can be determined several of them are still in service.

The commercial CFA design was patented by Hately and Kabbary in 1986, and Maurice Hately went on to produce a number of antennas for the amateur market using what he said was the same principle of PVS. These were the so-called Cross Field Loops (CFLs) for the amateur bands from 6m to 160m, a design which was also patented. The circuit diagram of the CFL antenna as shown in the publicly-available patent can be seen in **Fig 6.3**.

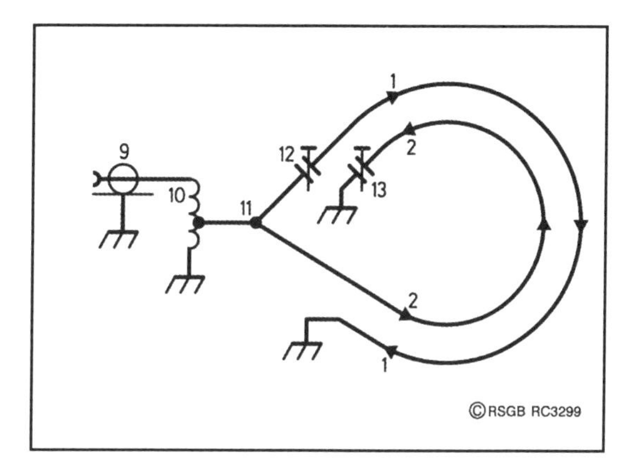

Fig 6.3: Circuit diagram of the CFL antenna as shown in the publicly-available GB Patent 2,330,695 (June 2002) and US Patent 6,025,813. Conductors 1, 2 may be coaxial in close proximity. 9 is the coaxial feeder socket. 10 is a matching auto-transformer. Capacitors 12 and 13 are said to be adjusted to be 45° ahead of and behind resonance.

The CFLs were very small – less than 1m in diameter for the 80m band. Steve Nichols, G0KYA, reviewed the 20m and 40m CFLs in the May 2002 *RadCom*.

The CFLs consisted of a loop attached to a sealed matching device, which GM3HAT insisted should not be opened. The 20m CFL included a 30cm loop of RG-58 coax, while the 40m antenna had a 40cm diameter loop made of copper tubing. In both cases the loop is held horizontally by a short plastic spreader. GM3HAT recommended using a length of 14m, or any multiple of 14m, of coax for the feeder.

At the time, GM3HAT gave this explanation of how the CFLs worked: "This antenna is the latest form of Crossed Field Antenna. The electromagnetic waves are created within the small 'Field Interaction Zone' around the two conductors of the loop. The new process of Poynting Vector Synthesis we have recently invented is the electrical dual of the system we used with success in the voltage stimulated forms of CFA Ref GB Patent 2 215 524 etc and papers in *Electronics World* March '89 and Dec '90 plus IBC Amsterdam '97 *IEE Conf* Publn 447 pp421-6. The CFL 7 is tuned internally to transmit 100W anywhere within the 40 metre amateur band. There are no user adjustments. The SWR when fed by 50Ω coax is less than 1.5:1 over the UK 7MHz band [then 7000 – 7100kHz – *Ed*]. If it is necessary to operate in the USA band an ATU will correct any small error."

G0KYA found that the 20m CFL exhibited a 1:1 SWR in the middle of the band, rising to 1.5:1 at either band edge. Mounted at about 20ft above ground, he found the CFL to be about 2 S-units down on a half-wave dipole, but on par with a loft-mounted magnetic loop antenna.

The 40m CFL performed rather better. It too had a 1:1 SWR from 7000 – 7050kHz, rising to 2:1 at 7100kHz and higher. UK and European signals were generally on par with a full-size dipole or, at worst, about 1 S-point less than the dipole. Although somewhat down on the dipole, G0KYA found he could work European stations at 59+ on the 40m CFL.

G0KYA also tested a multi-band version of the CFL but found that his other antennas always out-performed it by 2 S-points or more. Nevertheless, he concluded that the 40m loop worked well and it was certainly worthy of further investigation for those who might have no space for a full-size dipole or other 'conventional' sort of antenna.

Steve Nichols, G0KYA, with the Hately Cross Field Loops for 20m and 40m.

PYRAMID NVIS ANTENNA FOR 60M

When experimenting with antennas, it is the desire of most radio amateurs to have their antenna radiate at the lowest angle possible and thus provide the strongest signal at long distances. However, when the intention is to provide the *highest* possible angle of radiation, for short-range NVIS (Near Vertical Incidence Skywave) operation, a novel approach is required. One of the reasons UK amateurs were allocated frequencies around 60 metres (5MHz) was to investigate short-range communications using NVIS at times when the critical frequency is too low for good propagation at 7MHz and too high for fade-free propagation at 3.5MHz using NVIS.

John Pegler, G3ENI, developed a compact omni-directional antenna for use on 60 metres and designed specifically for NVIS purposes, radiating primarily in the vertical direction. It was reported by Pat Hawker, G3VA, in 'Technical Topics' (RadCom, December 2002). G3ENI wrote: "This antenna is ideal for the small garden, can be assembled easily for portable use, has low visual impact and can be located clear of other antennas. It consists of a half-wavelength of wire with the two arms bent in the form of triangles and assembled in the shape of a square-based pyramid as shown in **Fig 6.4**. When radiating, there is little reaction between the sloping wires that are at right angles, with orthogonal polarisation.

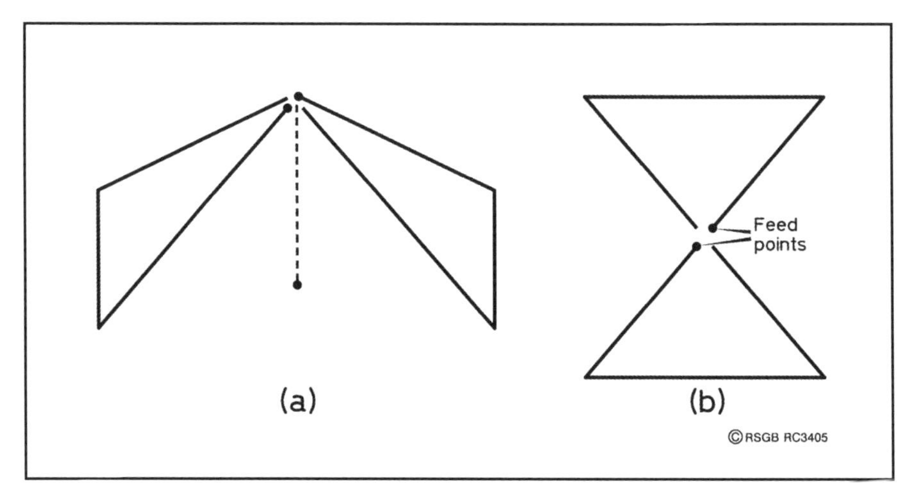

Fig 6.4: G3ENI's 'Pyramid' NVIS antenna for 5.4MHz. (a) General appearance. (b) Plan view. Dimensions for 5.4MHz in text.

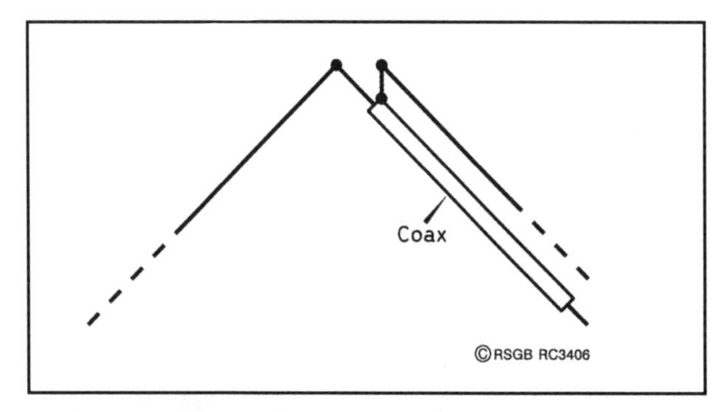

Fig 6.5: Sloping feedline to the Pyramid antenna. A balun could be used at the top of the pole if the coaxial cable is dropped vertically down the supporting pole.

"The apex is supported on a short insulated pole and the base points are secured to tent pegs or similar insulated ground anchors so that the horizontal sections are some 1 to 2ft above the ground. The feeder can either be led up and secured to the insulated pole or fed down one leg as shown in **Fig 6.5**, making sure that the coaxial cable outer is connected to the adjacent wire as shown.

"For 5.4MHz, the total length of wire required is about 88ft (26.84m), the sloping wires 13ft (4m), and horizontal wires 18ft (5.5m), for a mast height of 11ft (3.35m) which includes the extra 2ft to keep the horizontal wires off the ground. Note that with high power there will be quite high RF voltage on the horizontal wires at the low height of 2 to 3ft. For non-NVIS, there will be a loss of about 16dB compared to a half-wave dipole at 30ft."

2.4GHz 'PATCH' ANTENNA

This novel 7.5dBi gain antenna for experimenters was described in the November 2011 *RadCom* by John Heath, G7HIA (now sadly a Silent Key), who readily acknowledged that he based his design on dimensions and drawings provided by K3TZ (see www.qsl.net/k3tz).

G7HIA made the following modifications:

- He increased the thickness of the brass patch to make the antenna more rigid;
- and increased the spacing to 3.1mm so that M4 hex nuts, lying flat, would make convenient and accurate spacers during construction.

Patch antennas at these frequencies require accuracy and symmetry if they are to deliver the expected results. Consequently the home constructor is faced with two difficulties, the need for workshop skills and the availability of the right materials in such small quantities. Assuming you have the skills and access to materials, **Fig 6.6** shows the dimensions and constructional information.

The original design was for a patch giving left hand circular polarisation (LHCP) to be used as a dish feed. LHCP is used because when receiving right hand circular polarised satellite signals the polarisation will be inverted due to reflection off the dish. By simply turning the patch over, the antenna changes from LHCP to RHCP as shown in **Fig 6.7**. The antenna was originally designed for AO-51 satellite operation. This worked very well, as shown by the picture from Ivo, PA1IVO, who used it as part of his portable satellite system using the Arrow antenna for mode U/S (70cm uplink/2.4GHz downlink). AO-51 is not currently operational on 2.4GHz, but

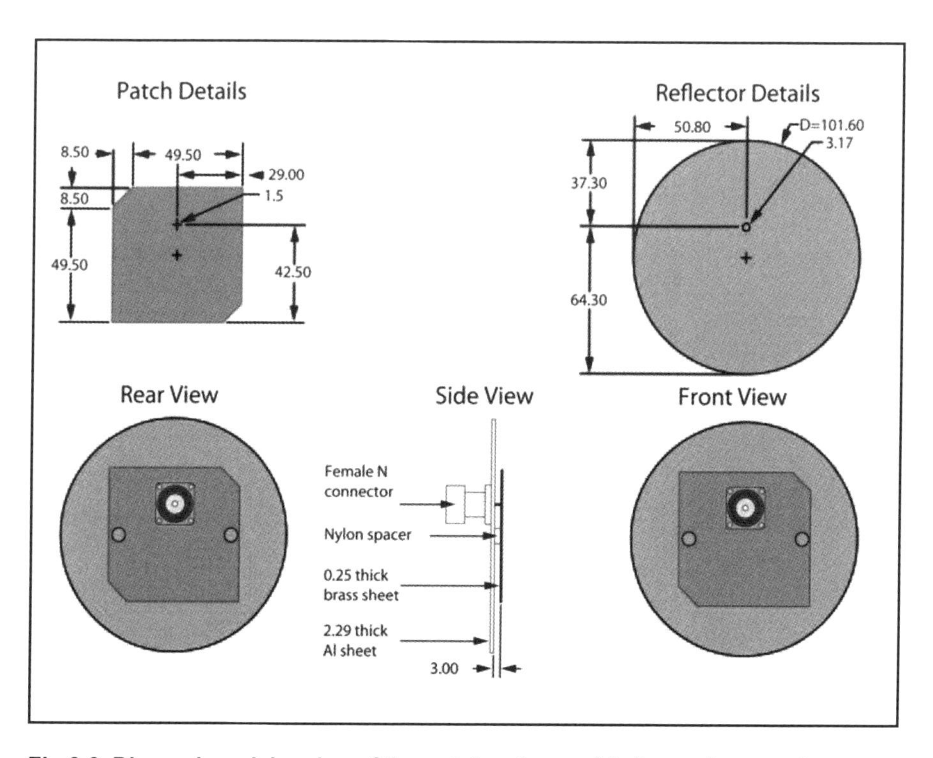

Fig 6.6: Dimensioned drawing of the patch antenna. All dimensions are in mm.

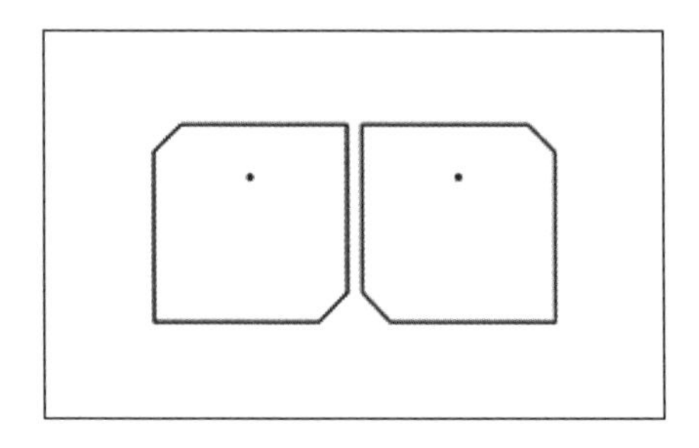

Fig 6.7: Orientation of the plate for left and right hand circular polarisation, left and right respectively, viewed from the front of the patch.

The 13cm patch antenna is quite compact and neat – especially when paired with a food container as a water-resistant radome. Photo by PA1IVO.

this antenna remains very useful for other 13cm activity.

Many experimenters have constructed patch antennas based on the K3TZ design and report good results; however, the question remains open as to whether the antenna produces truly circular polarisation. It is certainly 'good enough' for most amateur applications.

Note that the patch antenna is not waterproof. Any water between the patch and back plate will seriously degrade performance. G7HIA suggest a cover using a plastic jar or similar if the antenna is to be left outdoors. The photo above shows a plastic food container used as a neat radome by PA1IVO.

Test results on a completed antenna showed good impedance match and a return loss of -27.13dB at 2400MHz (see **Fig 6.8**).

G7HIA suggested some improvements to the published design: it could probably benefit from some additional mechanical support. A small nylon screw through the back plate and touching the back of the patch should work. The contact point should be at the centre of the patch where it will have least effect on the radiation characteristics. He acknowledges K3TZ for the inspiration and the workshop drawings; 7N1JVW who, he believed, provided the original design, modified by subsequent experimenters; David Bowman, G0MRF, who provided helpful advice during the project, and Howard Long, G6LVB, who provided the test results.

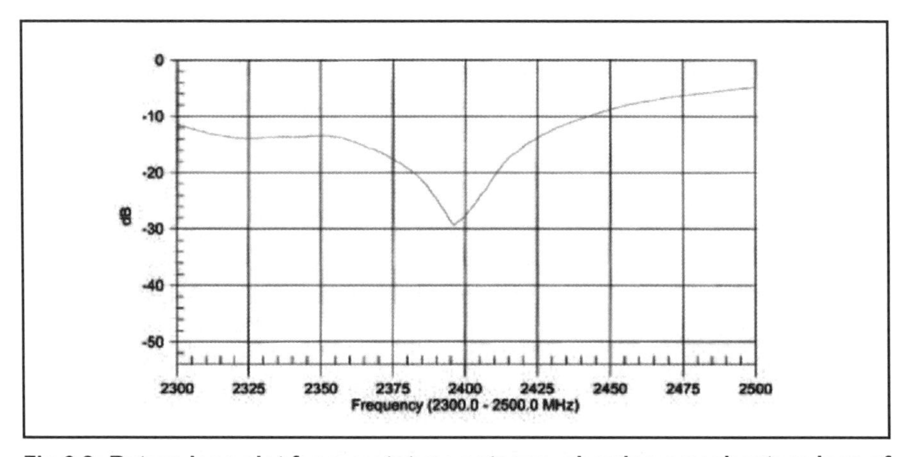

Fig 6.8: Return loss plot for a prototype antenna, showing a peak return loss of -29.37dB and a -20dB return loss bandwidth of about 29MHz. Courtesy G6LVB.

NOVEL ANTENNA FOR 472 – 479KHZ BAND

From microwaves to the other end of the spectrum. This 'antenna' is certainly 'novel', because it is not really an antenna at all! Roger Lapthorn, G3XBM, described it in the March 2013 *RadCom*. He wrote: "Like many, my home is not blessed with a large garden, so the mega-antennas used by some on the LF and MF bands are just not possible. Instead, small Marconi verticals and wire loop designs have been tried with varying degrees of success. In the past, I have used earth-electrode pairs for my VLF work, so I thought it was worth trying these on the new 472kHz band to see how such a system would perform. I was not expecting very good results but was amazed how well it worked, both on transmit and receive."

The earth-electrode pair consists of a pair of earth rods, in this case about 1m long copper earth stakes, driven into the soil about 20m apart in the garden at the back of G3XBM's house. An alternative arrangement is one earth rod at the far end of the garden and a connection to a metal water pipe inside the house, which in most cases finds its way electrically to ground as the water supply enters the property. (However, this may not be the case for more modern properties where water is supplied through plastic pipes. A continuity check between the 'far' earth stake and a metal water pipe should indicate up to about 150Ω if the pipe is nicely earthed; the resistance will be rather higher if the pipework is plastic.)

The output of the transceiver or transverter is connected directly to the two earth rods using ordinary PVC-covered wire. G3XBM used 32 x 0.2mm (1mm^2) wire. The current was measured with an in-line antenna current meter (A suitable circuit appears in the RSGB book *LF Today*, by Mike Dennison, G3XDV, and Jim Moritz, M0BMU). Depending on the impedance of any particular earth-electrode pair, some form of matching may be necessary.

The wires are simply laid out across the grass, as shown in **Fig 6.9**. It is important that the space between the two electrodes is not bridged by other earthed structures: the best results will be when they are far apart and 'in the clear', far from any other pipes, buried metalwork or cables.

To test the performance of this earth-electrode pair antenna, it was necessary to do some comparisons with other antennas. G3XBM used a small 9m high, top-loaded, Marconi vertical tuned to 472kHz using a fixed inductance near the top of the vertical section and additional inductance in series at the bottom to bring the whole antenna to resonance. He also tried an earth-electrode pair with one wire elevated, forming a 3-sided vertical loop.

To compare the field strength from each antenna he ran a series of *WSPR* beacon transmissions over a period of many hours with each antenna, making a record of the received signal-to-noise ratio logged on the Internet *WSPR* database by reporting stations with each antenna. To avoid any chance of one antenna disturbing the other, only one antenna was erected at a time. The output from the transmitter, an FT-817 with home-made transverter, was around 10W – although the radiated power is a fraction of this.

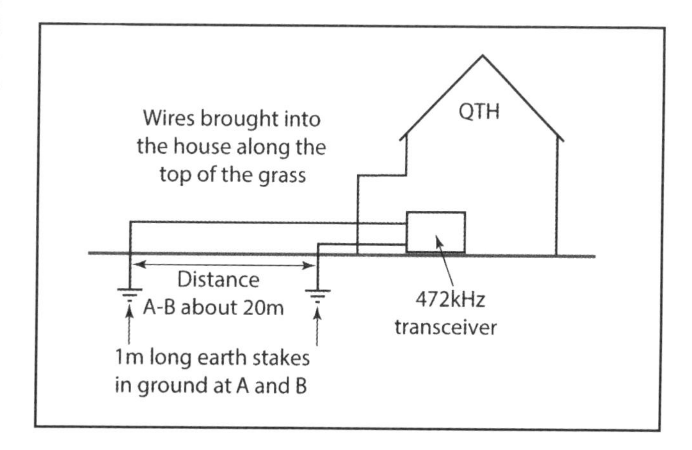

Fig 6.9: The basic G3XBM earth-electrode 'antenna' set-up.

Based on several days of observations of the G3XBM signal by stations up to 701km distant, it was established that the earth-electrode antenna is between 2 and 14dB down on the Marconi antenna, with worst

Reporter	SNR (dB)	Grid	km	Azimuth
DL-SWL	-28	JO52hp	701	83°
PI4THT	-30	JO32kf	448	88°
PA0A	-25	JO33de	417	73°
PA3ABK	-30	JO21it	306	98°
M0LMH	-30	IO93gx	223	329°
G0KTN	-27	IO81ti	210	242°
G4AGE	-25	IO93if	151	316°
G0MQW	-23	IO91ml	123	225°
M1GEO	-28	JO01cn	79	184°
G3ZJO	-20	IO92ng	79	270°
M0BMU	-30	IO91vr	69	210°
G0BPU	-23	JO02ob	67	110°
G7NKS	-16	IO92ub	46	240°
G8KNN	-11	JO02bf	12	248°
G4HJW	+7	JO02de	9	180°

Table 6.1: *WSPR* spots received by G3XBM when using <10mW EIRP from the earth-electrode antenna.

results from stations at right angles to the structure. For stations 'in line', the difference in performance was only a few dB. In other words, this antenna is not as good as the Marconi vertical described, but it still works pretty well and gets reports from a long way away (see **Table 6.1**). It is equally effective on receive, with good signals received from many stations across Western Europe on *WSPR* and on CW.

But how and why does it work? Although it is not possible to be certain, G3XBM believes the best explanation is that the earth-electrode pair acts as a kind of vertical loop in the ground. See **Fig 6.10**: current flows from one earth rod (A) into the ground and returns to the other rod (B) via a series of diffuse paths within the soil and rock beneath the structure. How far down and out the signal spreads will depend on soil chemistry and on the geology of the rock beneath. At G3XBM's location the ground is a light alkaline soil over chalk bedrock less than 2m below the surface. Instinctively one would expect the largest loop to be formed in low conductivity rocks and soils as the current has to 'spread out' more. In the limit, if the soil between the two earth rods was a perfect conductor, then all current would return directly from A to B and the loop would have no enclosed area.

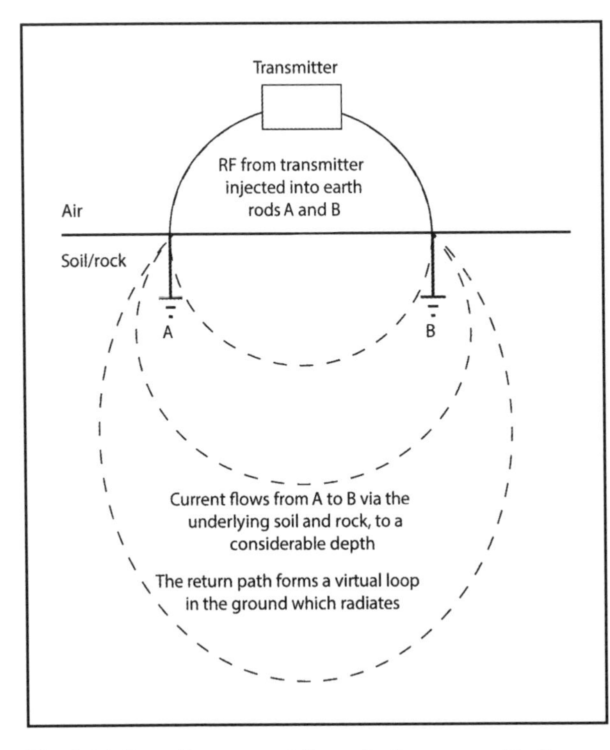

Fig 6.10: How the current flows in the ground to form a virtual vertical loop.

Using the signal strength report data and knowing the distance to the various listeners, it is possible to work out the effective isotropic radiated power (EIRP) used. Based on some well-proven formulae, Belgian amateur Rik Strobbe, OR7T, then calculated the effective size of the loop within the ground that would be needed to produce this EIRP. He calculated that the virtual loop in the ground was as large as 290 square metres – which is very big. Measurement errors mean this could be an order of magnitude out either way, but it has to be quite large to work as well as it does. In my case, the loss resistance of the earth-electrode system works out as being 66Ω (so I am able to directly connect from my rig into the earth-electrode antenna) and the radiation resistance is around 0.017Ω, to achieve the field strengths I do. Only a tiny fraction of the power from the transmitter is radiated, but this is also the case for other small antennas at LF and MF.

For information, the radiation resistance of a small loop Ra = $320*\pi^4*A^2/\lambda^4$ (ignoring the presence of any ground plane under the antenna) and $2*320*\pi^4*A^2/\lambda^4$ when this is included. A is the enclosed area of the loop in square metres.

It is possible to improve the performance of this structure by raising the height of the connecting wires, forming part of the loop in the air as well as in the ground. This is indeed how this antenna started out, until it was suggested that it should be tried with the wires laying on the ground in the grass: performance was almost unchanged and it is much easier to string a piece of wire across a lawn than elevate it up in the air. The small difference in performance as a result of elevating the 'above ground' part of the loop adds credence to the 'large loop in the ground' theory.

Clearly, most people will want to radiate as much power as possible, up to the legal limit of 5W EIRP on the 472 – 479kHz band. However, many are unable to erect large antennas and are prepared to accept a reduced EIRP, especially if the 'antenna' system becomes ridiculously simple. With the performance of the earth-electrode averaging just 8dB below that of G3XBM's 9m-high Marconi, he was still able to get *WSPR* reports from a large number of stations across Western Europe. The performance is certainly good enough for it to be used for semi-local CW QSOs out to at least 50 – 75km, even though the EIRP is only around 5mW from the 10W transmitter.

From his earlier work at 136kHz, G3XBM knows this structure is able to radiate a signal over a considerable distance, even with very low EIRP levels. At 136kHz, if one is prepared to accept the compromises, such an approach is a useful alternative when compared with a very large 'in the air' antenna with large, low loss, loading coils.

The earth-electrode pair can be a very useful alternative 'antenna' for the new 472 – 479kHz band. Although results will very much depend on local soil, rocks and the degree of metal clutter in the garden, it is certainly an antenna to try when larger antennas are not possible.

THE 'BALANCED LADDER' ANTENNA

This 'antenna' may be considered to be just a bit of fun, but it does illustrate a serious point: provided that RF current can be made to flow in a metal object, it is bound to radiate. In his 'Antennas' column in the April 2006 *RadCom*, Peter Dodd, G3LDO, wrote that by chance he came across the website for the K0S 'Strange Antenna Challenge' (www.strangeantennachallenge.com).

The idea of the challenge is simple. Construct an antenna from something that is not normally used as an antenna. Antenna materials used in some of the challenges have included such items as metal folding chairs, shopping trolleys, chicken wire, fences, ladders and trucks. Antennas constructed of wire or metal pipe or tubing (normal antenna materials) are *not* permitted. A feeder is allowed but efforts must be made to prevent the feeder from radiating using current chokes or baluns if using coax, or a balanced feed if using twin feeder.

Inspired by this and in preparation for the challenge, G3LDO constructed an 'antenna' from two step ladders and placed them on a garden table. The two ladders were fed as two elements of a dipole using a two-metre length of 400Ω ladder line feeder. The antenna exhibited a broad resonance at about 23.5MHz with a feed impedance of around 12Ω at resonance. The antenna was tried on several bands and loaded on all HF bands, 10MHz

Two step ladders set up for the 'Strange Antenna Challenge'.

The two ladders used by G3LDO in the 2006 'Strange Antenna Challenge'.

and up. Using an MFJ ATU or SEM Z-match and an Icom IC-706 on CW, G3LDO was surprised at how well it worked. At 100W, contacts were no more difficult than working with QRP (less than 5W) into a 'normal' 10m high multiband dipole antenna. In just over one hour he made five contacts around Europe and, on average, his reports were about two S-points down on the ones he gave out.

G3LDO then moved the antenna to the flat roof of his house extension, thereby increasing the height from 0.8m to 3.2m. The resonance was about the same and the feed impedance increased to 16Ω. This move improved the performance so that the average signal strength difference was less than one S-point. He even managed to work a couple of east-coast USA stations.

The 2006 'Strange Antenna Challenge' took place between 27 and 29 May and G3LDO decided to participate, using the two halves of the extension ladder laid out on the flat roof of the house extension about 10ft (3.2m) above the ground, as shown in the photograph. The antenna was tried on several bands and loaded on all the HF bands from 7MHz and up, when used with the diminutive MFJ-901B ATU.

So how did it work out? In its new location, the antenna worked much better and during intermittent operating sessions within the dates mentioned above, he made 19 contacts with 15 countries.

The Strange Antenna Challenge called for a fair amount of information exchange on the equipment, antennas and meaningful reports. With inter-State contacts in the USA this would be no problem. However, with inter-European contacts and operators not working in their native language, G3LDO's description of the ladder antenna was often met with "¿que?" or its equivalent. There were also some stations working 'contest-style' who gave the usual meaningless 599 reports.

Short skip contacts with Europe using 100W were fine, with the reports received similar to those given out. DX was a different matter: G3LDO called quite a few DX stations but generally was not heard. The exception was John, N1FOJ, in New Hampshire, who was using a beam and 800W. G3LDO received a 55 report and sent 59 to N1FOJ, and Peter says he felt that this was an honest report because the strange antenna challenge was discussed.

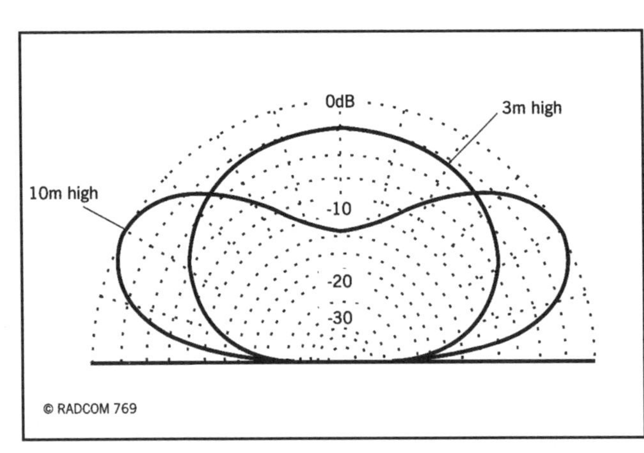

Fig 6.11: Polar diagrams of a short dipole (ladder) at 3m and 10m, which shows a 6dB improvement at 30°.

The interesting aspect of this exercise is not *what* the antenna is but *where* it is. By moving the antenna from one location to another the performance was improved no end. The lack of DX capability of this strange antenna was due to the fact that it was a horizontal antenna only 10ft high. Placing it 30ft high would have made a large difference to its potential DX capability, as shown in **Fig 6.11**.

Chapter 7
Commercial Antennas

MFJ WIRE LOOP TUNERS

If you are unable to put up an outdoor antenna for whatever reason, the American company MFJ has come up with a novel solution. MFJ manufactures a series of tuning units designed specifically to tune indoor loops of wire. Most conventional magnetic loops typically use rigid copper tubing as the radiating element. Using thinner diameter wire instead increases the losses but these losses can be reduced to some extent by making the wire loop larger than would typically be the case with a rigid copper loop for the same frequency band. MFJ recommends the use of 4mm diameter wire, but their tuners will also work with somewhat thinner wire.

The current (2015) models shown on the MFJ website (www.mfjenterprises.com) are the **MFJ-933,** the **MFJ-935B** and the **MFJ-936B**. The MFJ-933 is the 'budget' version and contains the tuning unit only. However, since the tuning of the loop is very sharp, a visual indication of when the unit is properly adjusted is virtually essential, and the MFJ-935B version provides this with a built-in antenna current meter. The MFJ-936B further adds an SWR bridge and power meter.

The circuit of the MFJ-935B is shown in **Fig 7.1**. VC1 adjusts the coupling from

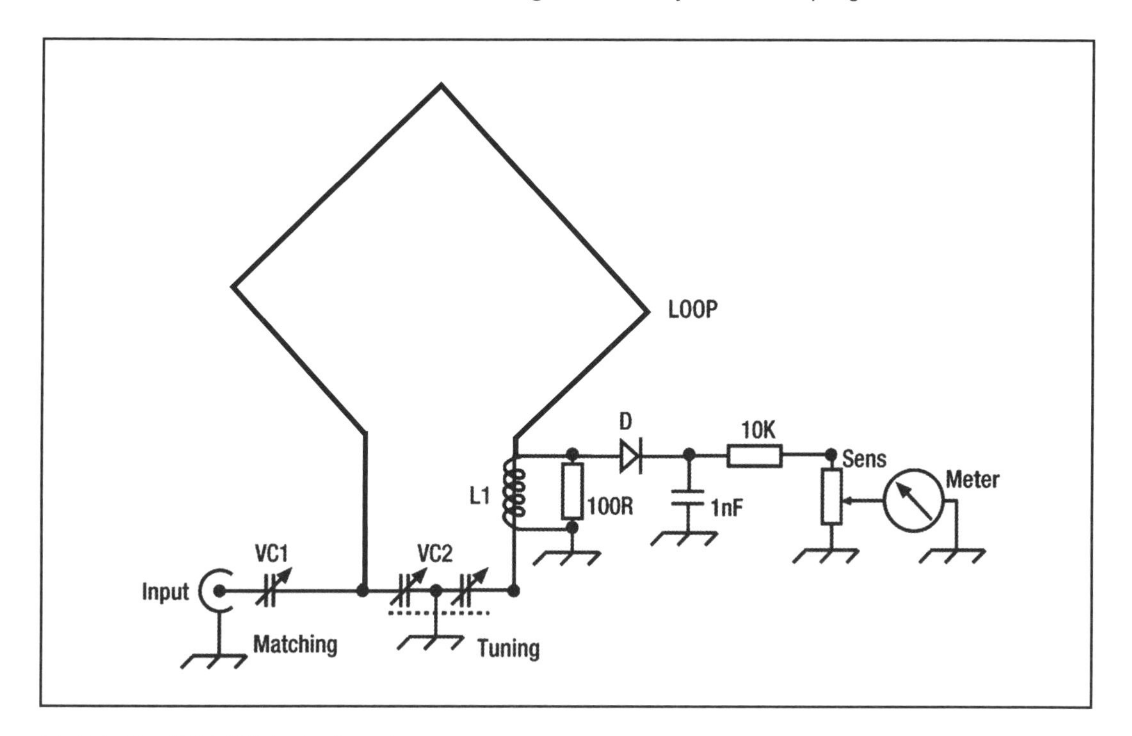

Fig 7.1: The MFJ-935B wire loop tuner.

Left: The MFJ-935B wire loop tuner. Note the plastic socket on top for attaching optional PVC spreaders to support the wire loop. Right: Showing the construction.

the transmitter to the loop and VC2 tunes it. VC2 is a butterfly capacitor with the centre grounded, so there is no voltage on the shaft. One of the output connections passes through a toroidal pickup coil, L1, which senses the current flowing in the loop. This is rectified and fed to a moving coil meter via a potentiometer, which indicates the relative value of that current.

The units will tune loops of approximately one-sixth of a wavelength to one-quarter of a wavelength long. The coverage of any one length of loop is about 0.75 of an octave, so different size loops must be used to cover the various bands. The units can operate from 80 to 10m, depending on the length of the wire loop used, and they are all rated at 150W.

A plastic socket is fitted to the top of the unit, on to which PVC 'spreaders' can be fitted and used to support wire loops for the 20, 17 and 15m, or the 17, 15, 12 and 10m bands. For best performance the loops should be configured to contain the largest symmetrical inside area. A circle is optimum and a square comes next. As the loop is distorted into a rectangle, so the efficiency drops. The PVC spreaders, which are an optional extra from MFJ (or you can easily make your own), allow the loops to be made into a proper square, or rather diamond, shape.

Several years ago I was able to test the MFJ-935B unit while working for the RSGB and living in South-East England. I immediately discovered that the location of the loop within the room made a big difference. Initial trials with the loops located in the middle of a room proved disappointing. Although many signals could be heard on the higher bands, it was difficult to make any contacts with the power levels attempted (50W or less). Results were much better when, instead of using the PVC spreaders, the antenna wire was placed around a window frame. Naturally the performance would be better still if the antenna was outdoors, but these units are designed specifically for those who are unable to put up any sort of outdoor antenna. Note that the wire loop must be connected directly to the tuning unit, and not via a length of feeder, and that you have to adjust the tuning unit by hand each time you change frequency by more than a few kilohertz).

When using power levels of 10 to 50 watts on SSB, good contacts were easily made around Europe on 20m SSB and around the UK and Western Europe on 40m while using a 7.92m (26ft) length of wire around a window frame – see photos. Compared with my full-size outdoor half-wave dipoles signals were generally about 1 to 2 S-points down. That might sound a lot, but in many cases the difference was almost undetectable, e.g. an S9 signal from Germany on the dipole was S8 on the MFJ loop tuner.

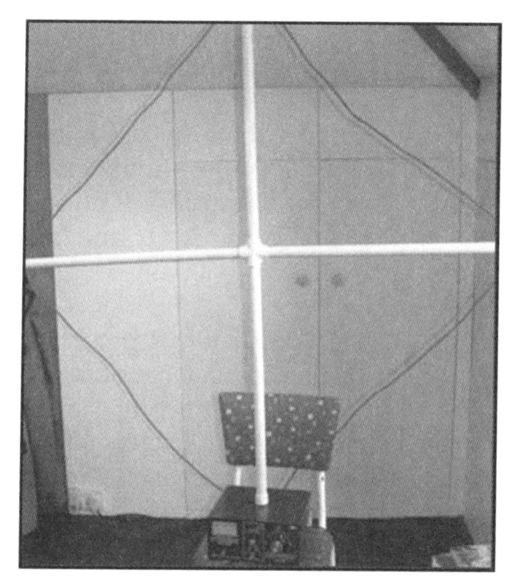

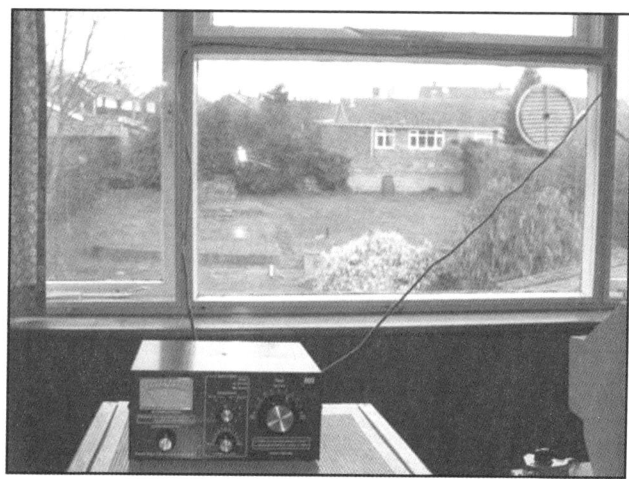

Left: With the loop in the middle of an upstairs room it proved difficult to make any contacts. Right: Wrapping a 7.92m (26ft) long wire around an upstairs window frame produced good results on 40m and 20m, with signals only 1 – 2 S-units down on well-sited outdoor half-wave dipoles.

Many amateurs do not have the space to put up outdoor antennas, particularly for the lower frequency bands, where a dipole for 40m is around 20m (nearly 66ft) long. Even the ubiquitous half-size G5RV is over 50ft long. Flat-dwellers may not be able to put up *any* external antenna at all. The MFJ wire loop tuners allow such people to get on the air and radiate a signal that can be within a couple of S-points of a well-sited outdoor dipole. Frequent travellers who often stay in hotels will also find a use for these units, as they will allow amateur radio operation to take place from locations that might otherwise have seemed to be impossible.

Safety note: MFJ quite correctly points out that high voltages and high currents exist on the wire loop while transmitting. Therefore, it is important that no-one is able to touch the loop when you are transmitting. A second precaution is associated with the field strength that exists near short loop transmitting antennas. It is imperative to keep a safe distance – MFJ suggests a minimum distance of two metres for all bands when operating at 100W, but a greater distance would be safer still.

THE BILAL ISOTRON ANTENNAS

Most antennas fall happily enough into the category of dipole, vertical, loop, beam or whatever. But as editor, I am glad I have a separate chapter for 'Commercial Antennas', because otherwise I am not sure in which chapter the Bilal Isotrons would have been included. They are none of the above, they are, in effect, radiating LC circuits, and that certainly makes them novel antennas in my book. There is nothing 'magical' about the design, though: the Isotrons are made up of two aluminium capacitor plates, top and bottom, with a coil mounted between them. This LC combination makes the antenna resonant on a given frequency.

The first thing to note is that this design makes these antennas *very* small. The 80m version is only 32 x 16 x 15in (81 x 41 x 38cm) in size. The 40m version is even smaller, at 22 x 16 x 15in (56 x 41 x 38cm).

The manual that comes with the Isotrons says that the antennas must be mounted on a *metal* pole, preferably earthed. It can be mounted in an attic or on a balcony, but even if you can't earth the antenna it should still be on a metal pole. The instructions say that if you mount them in the attic you could use the mains earth for

The 80m Isotron antenna.

your connection, but this seems like a recipe for RFI unless you are running QRP.

These unconventional antennas are available for the 160m–6m bands. Steve Nichols, G0KYA, tested the 80m and 40m versions in May 2010 and his review was published in the August 2010 *RadCom*.

The 40 and 80m Isotrons antennas as supplied can be set up independently or mounted together on a single mast and fed by a single length of coax. In the latter case, the interconnection between the two antennas is by a short length of supplied 300Ω slotted feeder.

An antenna less than three feet high is obviously attractive for amateurs who don't have the space for a dipole or similar full-size antenna. But can it possibly work? The following is extracted from the *RadCom* review by Steve Nichols, G0KYA:

"The antennas arrived in a very small cardboard box. Opening it revealed the two coils (one for the 80m version and one for the 40m). Also included were the aluminium top and bottom plates, the plastic / nylon insulators that hold the plates apart, plus all the hardware to assembly the antenna. The instructions came as a photocopied booklet, but were quite clear. It took about 30–45 minutes to assemble each antenna...

"You won't need much in the way of tools – I used a couple of 11mm spanners and a flat-head screwdriver. The antenna uses two aluminium rods to suspend the resonating coil between the top and bottom capacitor plates. The 80m version also has two small square aluminium tuning plates on rods that can be moved to tune the antenna to the part of the band you are interested in. You are advised not to fit these until the antenna is in position and you have found the natural resonant point, but given that it is designed to operate out of the box at around 3.950MHz (the US 75m band) you may as well fit one or both tuning rods from the start.

"Bilal recommends one tuning rod and hat if you wish to operate from 3.675 – 3.8MHz and two if you wish to operate from 3.5 – 3.675MHz. Once assembled the antenna is quite light (6lb / 2.7kg) and can easily be picked up with one hand. . .

"Once assembled I mounted it on a lightweight 18ft aluminium mast . . . Without the two tuning rods the antenna resonated at around 4.0MHz – way too high. But putting both tuning rods on, with the small aluminium capacity hats facing down, the SWR came down to 1.8:1 at 3.610MHz, using an earthed MFJ analyser. The 3:1 SWR bandwidth at this setting was 3.586 – 3.642MHz (56kHz). It is obviously best to set the antenna in the region of the band that is of most interest to you.

"Conditions on a May afternoon on 80m were not too good, but there were one or two SSB signals around. I compared the Isotron with my 132ft off-centre fed dipole (OCFD), which actually lies across the roof at about 30ft. I also have an 85ft W3EDP end-fed that also goes over the same roof. Both antennas perform about the same on 80m. I live in a typical suburban location and the noise level on 80m is usually S8–S9 all the time. I found the noise level on the Isotron about 3 S-points lower than on my normal antennas as it was positioned further away from the house. This made listening much easier. In terms of signal strength, signals were generally down about 1–2 S-points on the Isotron... The fall-off as you moved away from the resonant point was quite obvious, while at its resonant point the antenna was at times equal to my other antennas.

"A CW contact with Ray, G3ASG, showed that the antenna was OK until fading kicked in, then it became a bit of a struggle. Switching to the W3EDP made life a lot easier. That evening it was the same story. Contacts with F5VLO, G6NKL, M0KVA and G6UUR showed similar results to the afternoon. This isn't quite as bad as it sounds as most signals in the evening on 80m are often 59+10 to 20dB, so they become S8–S9 on the Isotron. However, if conditions are marginal the Isotron will lose out. It performed better on CW and PSK31 where absolute signal strength is not as critical.

"I passed the antenna to Roger, G3LDI, who mounted it at 45ft and compared it with a low 80m dipole at about 25ft. Roger found similar results to me – signal strengths were 10–20dB down with the Isotron and he found it noisier. He worked DO1DTA on 3635 getting a 59 report. Roger then switched to the dipole and received 59+20dB. Later he called CQ on the Isotron. G3OKA gave him 59, coming back to his first call too. Roger then switched to the dipole and he gave him 59+10dB.

"If you are looking for a replacement for an 80m dipole you will be disappointed, but if you have no other way of getting on the band it will work well for you, just make sure that you operate as close to its resonant point as possible for the best results."

Close up of the tuning rod.

THE 40m VERSION

G0KYA continues: "I then built the 40m version, which looks very similar, but is slightly shorter. The aluminium capacitor plates are also less wide than on the 80m version and it only has one tuning rod, not two. I fitted the tuning rod, complete with the small 1.5in square aluminium capacity hat and set it in the minimum capacitance position. I put the Isotron on the 18ft mast and found that it resonated out of the box at 7.050MHz with an SWR of 1.4:1. It also showed that it should be possible, by adjusting the tuning arm to resonate the antenna in the CW portion of the band. I then took the tuning arm off completely and found that the antenna resonated at 7.3MHz, so it looks like you do need the tuning arm on, at least for the UK allocation on 40m.

The 40m Isotron antenna.

"Tuning arm back on, but with no capacity hat, and I eventually managed to get the antenna resonant at 7.1MHz with an SWR of 1:1. The SWR at 7.000 and 7.200MHz was then 2.5:1. At the CW end of 40m the antenna was quite lively. Signals that were S9+10dB on the 132ft OCFD / W3EDP long wire were S8 on the Isotron, but then the centre of the dipole is 12ft higher. Some signals were only 1–2 S-points less on the Isotron, and quite a few were identical. In the SSB portion of the band, my first call was answered straight away by DL60DRC, a special event station in Germany. Other SSB signals were also either equal on the Isotron or down by no more than 1–2 S- points.

"The 40m Isotron didn't strike me as too much of a compromise. If you have no room for a 40m dipole the antenna will get you on the band. Again, if your interests are CW or PSK31 the antenna will serve you well. If you prefer SSB your signals are likely to be down by 1–2 S-points, but you will work the stronger stations.

"The lightweight 40m and 80m Isotrons allow you to get on the bands when you don't have room for a full-size dipole or long wire. Yes, signal strengths are likely to be down a little, but you will be able to operate. It pays to get the antennas as high as possible (a chimney would be ideal if you don't have a mast) and follow the installation instructions carefully to get the best results."

THE PRO ANTENNAS DUAL BEAM PRO

The Pro Antennas 'Dual Beam Pro' is a small lightweight antenna aimed at amateurs who want a multi-band antenna in as small and light a package as possible. The antenna is effectively a *non-resonant* dipole with capacity hat end loading. It has been specifically designed to be non-resonant on any of the amateur bands and uses a balun / impedance transformer at the feed point to lower the resultant SWR to a more manageable level.

The Dual Beam Pro mounted on domestic TV aerial gable wall fixings.

The Dual Beam Pro was reviewed by Steve Nichols, G0KYA, in the May 2011 *RadCom*. However, since then the design has been changed and the 'New Dual Beam Pro' now covers no fewer than eight bands (or even nine if you include 11m CB) – i.e. 6, 10, 11, 12, 15, 17, 20, 30 and 40 metres – and its power rating has been increased to 1000 watts PEP on 6 to 20m (600 watts PEP on the 30 and 40m bands).

The antenna looks like a giant H on its back: the main element is 5.0m long while the two end elements are 2.5m long. Aerospace alloys are used throughout, with excellent corrosion resistance and non-corrosive stainless steel fixings. The Dual Beam Pro comes supplied with a heavy gauge steel galvanised mast head support bracket for fitting to a 1.5 – 2in support mast. As the antenna is bi-directional it requires only 180° of rotation and because it is lightweight (4kg, including the support bracket and matching transformer) the use of standard domestic TV aerial chimney stack brackets or gable wall fixings are sufficient.

The non-resonant nature of the antenna means that an ATU is required on all bands. Some transceivers' internal ATUs may be able to handle the SWR on some bands, but a wide-range external ATU is probably to be recommended in most cases. The antenna is small and therefore the manufacturer acknowledges that "40m efficiency is down when compared to the other bands".

The designer, Carl Kidd, G4GTW, explained the thinking behind the design. With a conventional half-wave dipole you get significant nulls off the ends. This means that when fixed in any one direction you will find that signals being received 'off the ends' will be down considerably – perhaps by up to 3 or 4 S-points. Plus, to get five band operation (10 – 20m) you normally need either traps, parallel-fed radiators or some other form of matching to get the SWR down to 1:1 on each of the bands. The result can be quite a heavy, complex antenna that still needs to be rotated to give 360° coverage. With the Dual Beam Pro, its non-resonant nature and impedance transformer means that you can find a match more easily, the construction becomes much simpler and the antenna much lighter. Also, the capacity hat end loading makes the overall length shorter. At higher radiation angles the Dual Beam Pro starts to become more omni-directional when mounted at a height of, say, 10m. But at take-off angles less than 10° (such as needed for DX) you can see the nulls off the ends of the antenna quite clearly. These can easily amount to 10 – 48dB (2 – 8 S-points) depending on the band and height above ground. Therefore, for best results the antenna should be rotated.

So much for the facts and figures. But how does the antenna perform? The following is extracted from the 2011 review by Steve Nichols, G0KYA:

"The Dual Beam Pro arrived in two boxes – one long tube containing the radiating elements and a smaller box with the balun and mounting hardware. I built and tested the antenna at the location of Chris Danby, G0DWV, where he has an extensive antenna farm and a trailer-mounted Versatower that was used for the Dual Beam Pro. Using Chris's set up we were able to compare it with sloping / horizontal dipoles, a doublet and a G5RV all suspended at about 50ft – and even a Force 12 beam at 80ft. The Dual Beam Pro was mounted lower than the other antennas and about 100ft away to minimise interaction.

The balun and mounting hardware.

"The hardware includes stainless steel fittings and a galvanised mast head bracket. The centre support insulator is solid GRP rod that provides good structural strength together with very low moisture absorption characteristics.

It took less than one hour to assemble the antenna.

Frequency (MHz)	Measured SWR
7.100	20.9:1
10.120	10.6:1
14.000	4.7:1
14.350	4.3:1
18.070 – 18.168	3:1
21.000 – 21.450	2.8:1
24.930	2.6:1
28.000	2.6:1
29.700	3.2:1

Table 7.1: Measured SWR readings of the Pro Antennas Dual Beam Pro (at 25ft above ground and at the end of 100ft of RG-213 coax).

"It took less than one hour to assemble the antenna and you need little more than a couple of spanners – it is very easy to build and the instructions are clear. Carl recommends that you feed the antenna with at least 20m of coax, which helps reduce the SWR that the rig and ATU will see. The result is that most internal ATUs will be able to match the antenna down to an SWR of 1:1 [on the bands from 10 – 20m – *Ed*]. In reality, with the antenna on a Versatower at about 25ft and fed with 100ft (30m) of brand new RG-213 coax I found that the resultant SWR was below 3:1 on most bands, rising to a maximum of 4.7:1 on 14.000MHz (see **Table 7.X**).

"My own rig's internal ATU will tune antennas with an SWR of up to about 10:1, but I know that other manufacturer's will only manage about 3:1, so you might have to use an external ATU if you have problems finding a match on some bands.

"On hooking the Dual Beam Pro up to Chris's station, testing could be started in earnest. The HF bands were humming with the solar flux at 155. First impressions were that the antenna is quiet, noise-wise. This is probably as a result of it being horizontally polarised and balun-fed. In a noisy suburban neighbourhood this could be a major boon. The next impression was that this is no compromise antenna. It heard better than Chris's doublet and G5RV and was roughly equal to resonant half-wave dipoles (as you would expect). It was also electrically quieter than the G5RV, doublet and a half-wave sloper on 17m suspended with the top at 55ft.

"On tests with the SV5TEN beacon on 28.180MHz we found that rotating the Dual Beam Pro so that it was end on rather than side on resulted in the signal strength dropping by about 2 to 3 S-points. This was about what my modelling [with the *MMANA-GAL* antenna modelling software – *Ed*] predicted and shows that for best results the antenna should ideally be fitted on a rotator. Its lightweight construction, however, means that a smaller, less expensive, rotator should be fine.

"A list of DX worked and heard doesn't really tell you about the antenna's performance, but in tests with VU2DSI in Mumbai, India on 10m; VK4JUZ in Australia and W4UWC in Knoxville, Tennessee on 17m; and BA3AO in China on 20m, the Dual Beam Pro bettered all the afore-mentioned antennas in terms of the signal-to-noise ratio and overall signal strength. Only a Force 12 beam at 80ft performed better, usually beating the Dual Beam Pro by about three S-points (as you would expect). Back-to-back tests with a station in Sweden on 17m SSB confirmed that it was either equal to or better than all the other wire antennas by about 1 S-point. A further week of testing by Chris confirmed that is a very quiet, usable antenna that performs well on all bands 10 – 20m.

"I must admit we were both surprised as conventional wisdom would have it that the design should result in mismatched losses due to a higher SWR on the feedline and potential losses in the balun. In reality, this wasn't borne out and perhaps proves that a rotatable, low-noise dipole-like antenna (even if non-resonant), can give better performance than a fixed wire antenna like a dipole, doublet, G5RV or noisy vertical. The antenna can also be used on 30m and 40m with reduced performance. It is down on a dipole on these bands due to its small size...

"It isn't a directional beam (with gain) in the traditional sense, but then it doesn't pretend to be, weigh as much or cost as much. What you do have is an electrically-quiet, simply-constructed antenna that can be used on 10 – 20m with little fuss. You can rotate it to peak signals or null out the ones you don't want. You will need a good internal or an external ATU and it is best fed with quality coax like RG-213 rather than the thinner RG-58. Carl suggests Mini RG-8 [aka RG-8X or just 'Mini-8' – *Ed*] as a good compromise for cable lengths between 20 and 30m.

A trailer-mounted Versatower was used to support the Dual Beam Pro for the tests.

"In conclusion, both Chris and I started the tests thinking that the antenna would be a compromise. In the end we were both very impressed – and when it comes to antenna testing that doesn't happen very often."

The *New* Dual Beam Pro costs £239 in January 2015 and is available direct from the UK manufacturer's website at www.proantennas.co.uk

THE SUPER ANTENNA YP-3 PORTABLE YAGI

Why is a Yagi in a book about 'novel' antennas? In this case it is not so much the design that is novel, but rather its implementation. The YP-3 is the ultimate 'beam in a bag': it's not a multi-band antenna as such, but rather a kit of parts from which you can construct a 3-element Yagi on any of the bands from 6 to 20 metres. What other HF / VHF Yagi can be carried around in a bag only 3ft (less than 1m) long and weighing only 11lb (5kg)? *That* is why it qualifies as a novel antenna.

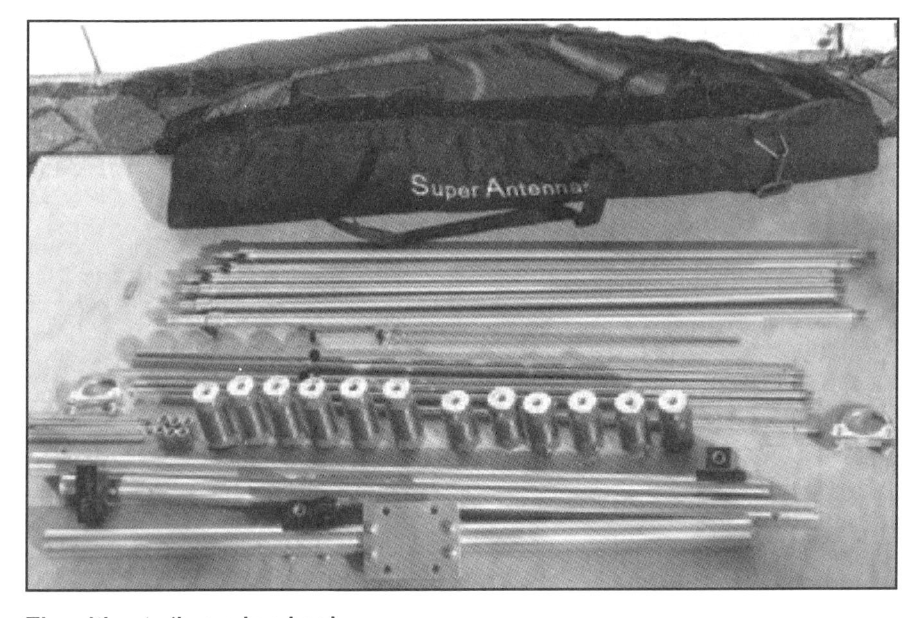

The ultimate 'beam in a bag'.

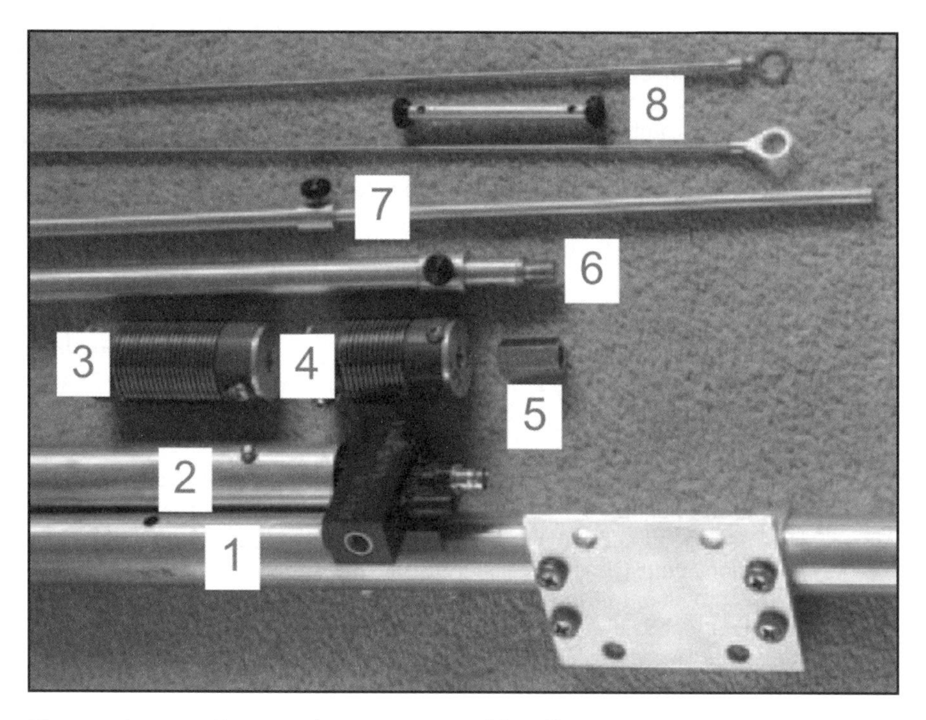

The numbers on the constituent parts are identified in the text.

The YP-3 was reviewed in the May 2009 *RadCom* by Peter Dodd, G3LDO. The kit of parts allows the construction of a 3-element Yagi using loading coils and with a maximum element span of 18ft 4in and a boom length of 10ft. On 6m and 10m bands full-size 3-element Yagis can be made, without the use of the loading coils.

The method of antenna construction can be seen by referring to the photo above. The centre section of the boom **(1)** comes with the driven element centre insulator and the boom-to-mast clamping plate already fitted. Two boom extensions **(2)** with the Reflector and Director centre insulators fixed at their ends telescope into the centre section (only one extension is shown in the photo). Spring loaded spigots in **(2)** locate into holes in **(1)** to lock the extensions in place. Additional holes in the centre section allow the boom length to be shortened for the 50MHz band. The centre sections of all the elements are made from 30in lengths of 5/8in aluminium tube into which a section of 1/2in tubing is telescoped **(6)**. The loading coils **(3)** or **(4)** are threaded on to the ends of **(6)**. Coil **(3)**, (7.52μH) is used only on the 20m antenna while coils **(4)** (3.62μH) are used on the 17, 15 and 12m antennas. The 10m antenna does not use loading coils and coupler **(5)** is used in place of the coils. Finally, the adjustable element end section **(7)** is threaded into the other end of the coils. The 6m antenna is made up of element sections **(6)** only. A hairpin matching device is used to match the feeder to the antenna. Each 1/8in brass hairpin rod has a 3/8in ring lug at one end **(8)**. The threaded butt of **(6)** is passed through the hairpin lug before is screwed into the driven element centre insulator; this is done on both sides of the driven element insulator.

The coax feed and hairpin matching arrangement.

The hairpin shorting bar is attached and fixed via the opposite ends of the hairpin rods as shown in the photo (left). The position of the shorting bar on the hairpin rods for any specific antenna is described in the instructions.

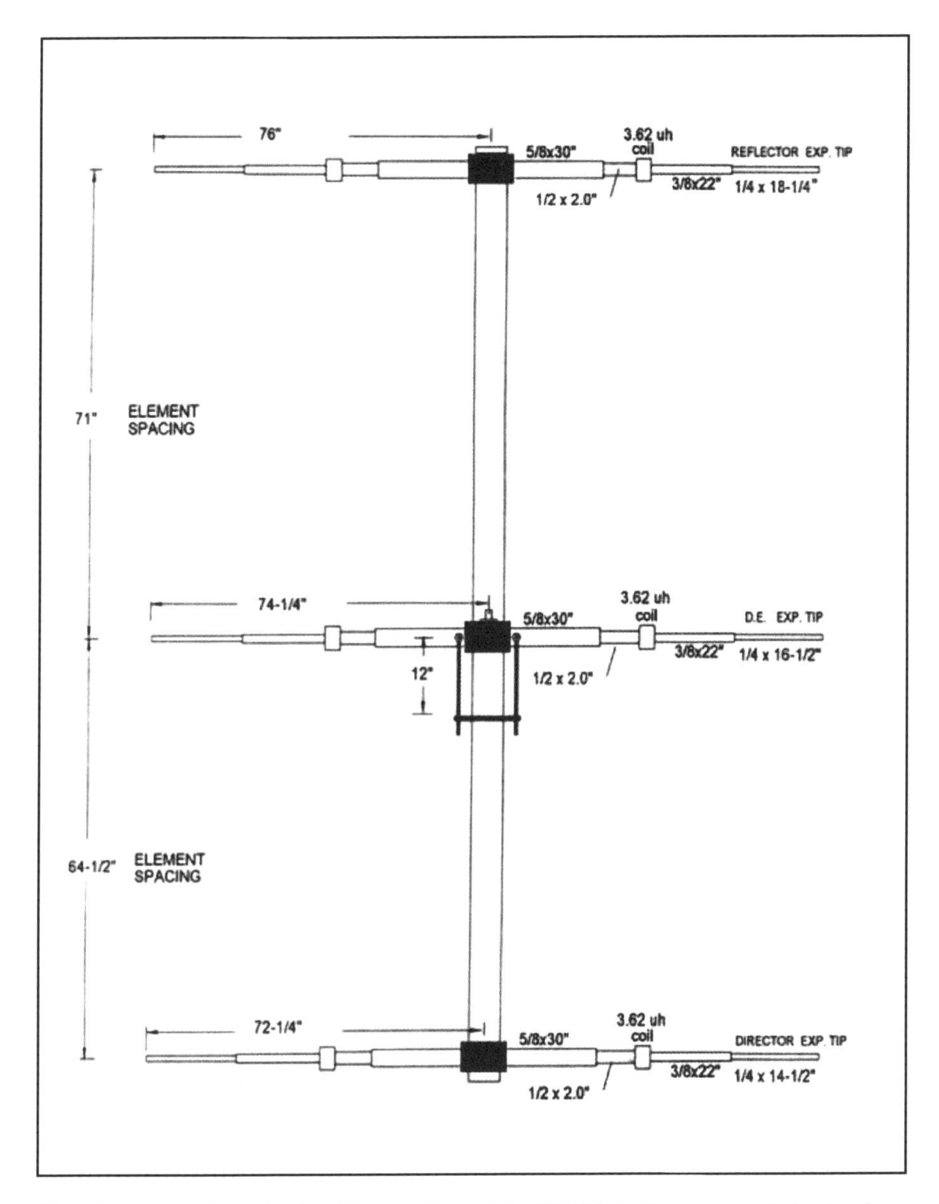

Fig 7.2: Dimensions for the 15m version of the YP-3 (all dimensions in inches).

The instructions specify exactly the dimensions of the element sections for any particular band antenna and a measuring tape, calibrated in inches, was supplied with the kit. I chose a 15m antenna for the test and the dimensions are shown in **Fig 7.2.**

Setting up the element sections is easy and the only tool required is a ½in spanner to fix the antenna to the mast. As you can see in the photo opposite the telescoping element sections are locked in position with thumb-screws. The coax connection to the driven element is with a BNC plug.

It is one thing to build horizontal beam antenna but quite another to raise it to a height that is effective in a portable environment. This may not be a problem to a radio club with the resources of a mast normally used on field days. For the single operator the simplest arrangement is a fold-over mast with an arrangement that

Suitable fold-over anchorage for a temporary portable mast.

uses a secure and stable mechanism as shown in the photo (left). The 15m version of the YP-3 is shown fixed to a portable fold-over mast in the middle photo.

The 15m and 6m versions were selected for test, to see if they performed as described in the instructions. The 15m version was raised and as you can see in the photo bottom left it looks good. A directional test was conducted with a local amateur G3CCX. The measured front-to-back was around 12dB. The measured SWR plot was very close to that provided in the instructions as shown in **Fig 7.3**. You will also notice that a gain figure of 9dBi is also given in **Fig 7.3** and the instructions say that this gain is "dBi over ground". However, the height of the antenna above ground, ground type and elevation angle of maximum radiation is not specified.

15m version of the YP-3 fixed to a fold over mast ready for raising.

On 10m, the YP-3 is a full sized monoband three element Yagi, which according to *The ARRL Antenna Book*, has a *free-space* gain of around 8dBi but is quoted in the instructions of having a *ground* gain of 13dBi.

An *EZNEC* model of the YP-3 implies that the performance figures have been obtained through modelling the antenna(s) 20ft (6m) over ground with a dielectric constant of 13 and a conductivity of 0.006S/m (i.e. average ground). In this model the gain of 13dBi on 15m was the same as quoted in the instructions and occurred at an elevation angle of 30°.

The 6m version is constructed from the boom and element centre sections – parts **(1)**, **(2)** and **(3)** in the earlier photograph – and is a full sized 3-element optimised spaced Yagi. As you might expect this antenna is quite broad-banded and I measured its 2:1 SWR bandwidth as 48 to 52.8MHz. It should also have a free-space gain better than 8dBi; in the instructions it is quoted as having a ground gain of 13.2dBi.

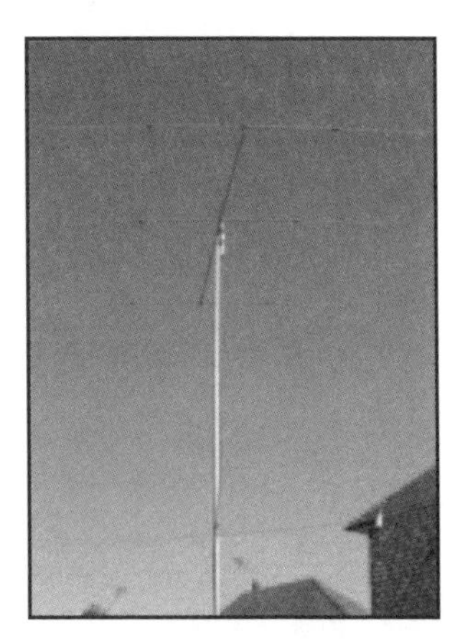

15m version of the YP-3 in the raised position, 6.5m high.

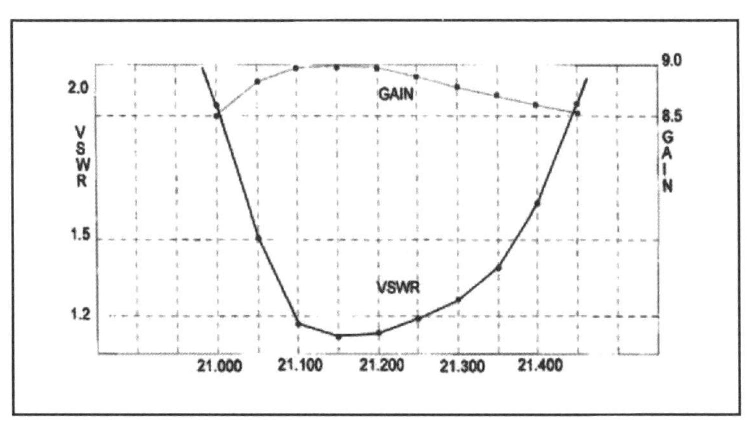

Fig 7.3: SWR and gain quoted in the instructions of the 15m YP-3.

The 6m version of the YP-3 is shown in the photo to the right.

This kit is easy to assemble and weighs only 11lbs (5kg). However, it requires a push-up or fold-over mast at least 6m (20ft) tall in order to realise the benefits of a horizontally polarised gain antenna.

The YP-3 supersedes an earlier 2-element model called the YP-2. This antenna was used in a 'shoot out', an event in the USA where antennas are compared with a reference antenna on a test range. This shoot out was done on 14.1MHz by the 'HF Portable Group'. You can read the conditions under which these tests were conducted at hfpack.com/antennas. The reference antenna for the horizontal group was a full size dipole made from 6AWG copper braided wire on a fibreglass support. The YP-2 registered a gain of +2.85dBd and a front-to-back of 18.64dB. Added to the results was a note which read "This is the highest gain ever tested in the HF pack antenna shoot outs". It is probable that the gain of the YP-3 is at least 2dB greater.

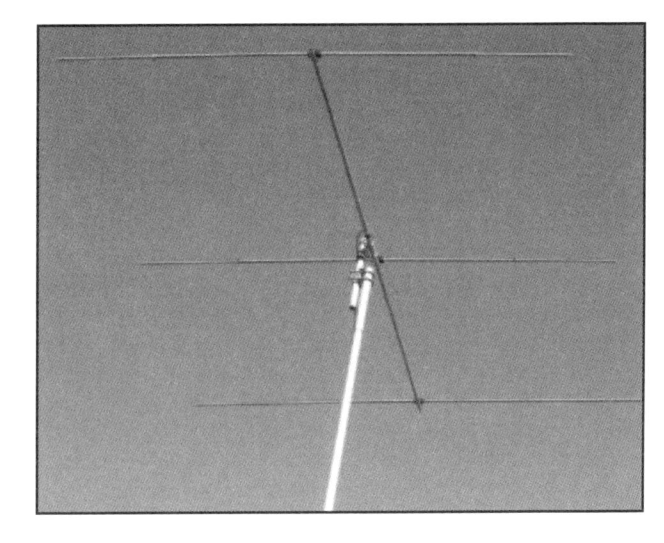

6m version of the YP-3: a full-size optimally-spaced 3-element Yagi.

The Super Antenna YP-3 is available from the Super Antenna website at newsuperantenna.com at a cost of US $469.95 plus P&P.

WELLBROOK ALA1530 RECEIVE ANTENNA

Sometimes it is difficult to know in which chapter an antenna should be included. The Welbrook ALA1530 *is* a commercially-made antenna, but it is also a receive-only antenna so, with the particular requirements of those wanting a low-noise receive antenna, it would be equally at home in Chapter 5. Noise from switch mode power supplies, plasma TVs, broadband over power line devices and much more is getting to be a big problem in many areas, so much so that some amateurs have even closed down altogether rather than face the battle against local noise.

So anything that can alleviate noise is to be welcomed, and the Wellbrook ALA1530 receive antenna has gained a world-wide reputation among short-wave listeners interested in the LF and MF part of the spectrum and amateur low-band DXers alike. What is novel about the Wellbrook ALA1530 is that it is an *untuned* loop with a built-in wideband preamplifier, so unlike most magnetic loops, and particularly those used for transmitting, it does not need to be retuned as you change frequency. It is designed to work from 50kHz to 30MHz and has been redesigned to increase the long wave and medium wave gain by approximately 10dB and 3dB respectively.

The antenna was reviewed by Steve Nichols, G0KYA, in the January 2012 *RadCom*.

How the loop arrived, packaged for transit.

The whole antenna comes through the post in one large package. When you unpack it there is little to do,

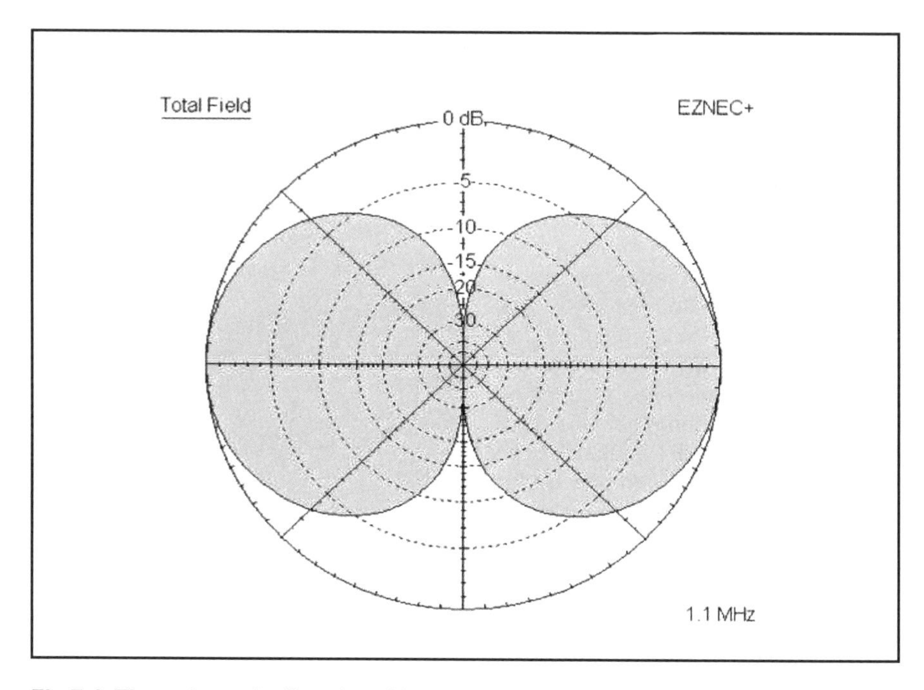

Fig 7.4: The antenna is directional in the plane of the loop.

other than fit the mounting flange and short mounting tube if necessary. The antenna comes with its own power supply interface and a small regulated PSU. The interface feeds 12V at 150mA via the coax to the BNC connector on the antenna (maximum recommended length 100m). A one metre lead fitted with a PL259 goes from the interface to your receiver or transceiver.

The loop is one metre in diameter and the wideband preamplifier is built into a plastic box at the base. The preamp is embedded in epoxy resin to help with weather-proofing and mechanical strength.

The antenna can be mounted directly to a piece of wood or other non-conducting surface, or the supplied aluminium mounting flange and short aluminium tube can be used to mount it on a rotator or mast. Wellbrook recommends the use of a rotator as the antenna is directional in the plane of the loop – see **Fig 7.4** – it has significant rejection off the sides, in the order of 35dB, that can be used to null out local interference or interfering stations.

Wellbrook recommends that it should be positioned approximately 5m away from buildings, metal objects and sources of interference. If using it as a receiving antenna in conjunction with a transmitting antenna you should keep them as far apart as possible. The company suggests that you can mount the antenna at ground level and G0KYA's tests were done with it on a short four-foot aluminium pole. It was fed with about 20m of Mini-8 (RG-8X) 50Ω coax. The antenna may be mounted higher if desired: this may improve HF performance, but is unlikely to help with LF / MF reception. There is also nothing to stop you installing it in your loft although it is unlikely to fit through the access hatch and the loft space is *not* the best place for low noise reception. To get around the first problem Wellbrook offers the LA5030 semi-rigid loop for indoor use. This will fit through a loft opening and costs exactly the same as the ALA1530.

For the reception tests G0KYA used Icom IC-7400 and IC-756 Pro 3 receivers which he states are somewhat insensitive on LF and MF, but it was the overall comparison with his wire antennas that was of interest. He started around 70kHz, where the Wellbrook brought in time signal stations that were virtually inaudible on a 100ft doublet. Moving up to the long wave broadcast band, many strong signals were found during daylight, including 153kHz (Deutschlandfunk), 162kHz (France), 183kHz (Saarlouis), 198kHz (BBC R4) and others which were generally clearer than on his wire antennas.

Further up the bands a host of non-directional beacons (NDBs) from around Europe were heard, such as 387kHz ING in St Inglevert, France and 395kHz OA, in Schiphol, Netherlands.

On the medium wave band the ALA1530 easily received distant stations, even in broad daylight. BBC Radio Scotland (810kHz) was perfectly audible in Norfolk. The directional effects of the loop were apparent when tuned to BBC Radio Wales on 882kHz from Washford, Somerset, which initially was inaudible. Rotating the antenna from NW / SE orientation to SW / NE made BBC Radio Scotland disappear and BBC Radio Wales appear (at a very clear S5). This shows how the directional capabilities of the loop can be used (on ground wave and low-angle signals) to null out interfering signals. On higher-angle signals it tends to be more omni-directional.

Further afield, WWZN in Boston, Massachusetts on 1510kHz was heard easily at 0330UTC in late August with the loop orientated NW / SE, followed by CFRB Toronto on 1010kHz and WWKB Buffalo, New York on 1520kHz. There were traces of these stations on the doublet, but nothing more.

The loop does not offer *stronger* received signals than a conventional wire antenna but it generally offers a much better signal-to-noise ratio, making weak signals much clearer. There were very few stations that were audible on the ALA1530 that were inaudible on the wire antennas, but the quality of the signals was improved, often dramatically on very weak signals with the loop. It became perfectly possible to listen to distant AM stations as if they were locals.

The ALA1530 under test at G3MPN.

On HF, the Welbrook loop received pretty much anything that G0KYA's wire antennas could. Interference from BT Vision PLT adaptors (which plagues the short wave broadcast bands, such as 19m and 25m at G0KYA's location) was still audible on the Wellbrook loop, but careful orientation reduced it dramatically.

The problem of noise starts to vanish as you head higher in frequency and the benefits of having a separate receive antenna on, say, 18MHz or 21MHz are not as apparent as having one on 160 or 80m. G0KYA found that weak, but perfectly audible, signals on 17m on his wire antennas were also weak, perfectly audible signals on the Wellbrook. In other words, if you are an amateur (as opposed to an SWL) there is little to be gained in using the ALA1530 as a separate receive antenna on the higher bands. However, go lower in frequency and it is a different story.

On 160m the Wellbrook turned a noisy S9 mess on a wire antenna into a perfectly quiet band. During daylight in September, a German SSB station could be heard on 1850kHz on the loop that was totally inaudible on the wire antennas. It was a similar story on many G stations around the UK on 160m. At 0350UTC, W0FLS, K4EJQ, K0ONF and K2JO were heard on the loop on 160m – admittedly weak, but there was no chance of hearing them at all on the wires. Listening to topband on the loop was

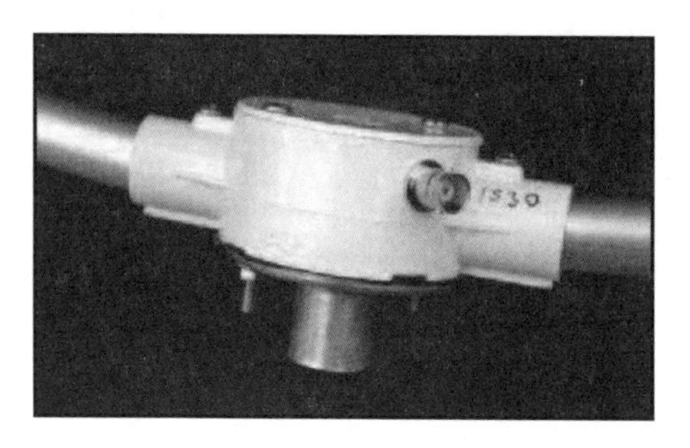

The feedpoint of the loop antenna.

an absolute pleasure – no noise, just pure CW signals. On 80m weak CW stations in the Netherlands could be heard clearly during the day on the Wellbrook that were barely audible on the wire antennas. The ALA1530 made them a lot easier to listen to.

The loop was then offered to David, G3MPN, and Roger, G3LDI, to test. Both stations use high, long doublets. Their results were broadly similar and both noted a marked reduction in noise on 160m – G3LDI said that readability is definitely improved by cutting out the noise, but a longer winter test might be necessary to draw firm conclusions in terms of overall S/N ratio.

For short wave listeners looking for a single, small antenna to cover everything you may ever wish to listen to, the Wellbrook ALA1530 could prove very useful. If you are an amateur looking for a receive antenna to augment your existing set-up, the ALA1530 will also fit the bill admirably, with the greatest benefits being on 160m and 80m, especially if you suffer from local noise on the lower bands. If you can locate the Wellbrook well away from your house you will notice a big difference and you are likely to be able to hear weak signals that are inaudible on wire antennas. To avoid damage it should be mounted at least 20ft away from transmitting antennas if running 100W, or 30ft away for 400W transmitters. It is easy to connect the ALA1530 to transceivers with a separate receive antenna input.

To get the most out of this antenna it is highly recommended that you use a rotator, otherwise you may miss out on a lot of signals and you will also miss the ability to null out interference.

The ALA1530 can be ordered from Wellbrook's website at www.wellbrook.uk.com

INNOVANTENNAS 9-ELEMENT 2M LFA YAGI

The term Loop Fed Array (LFA) describes a method of feeding Yagis that has been developed by Justin Johnson, G0KSC, of the UK firm InnovAntennas. In place of the usual dipole driven element, the LFA uses a rectangular shaped, full-wave loop driven element that is laid flat on the boom between, and in-line with, the parasitic elements (see photo). The smaller end sections of the loop, which run parallel to the boom, are 180° out-of-phase with each other: each side cancels the other out

The 9-element 2m InnovAntenna on test.

and so minimum radiation occurs, leading to highly suppressed side lobes and side-on signal rejection. InnovAntennas also claim that their antenna offers wide-band performance, unlike some other antennas that have their best front-to-back ratio at one end of the band and their best gain figure at the other end.

InnovAntennas offers a wide range of antennas for bands from 10MHz to 432MHz and their 2m 9-element LFA Yagi was reviewed by Steve Nichols, G0KYA, in the March 2012 *RadCom*.

The antenna comes well packed and, when constructed, is 4.43m long. On opening the box you are met with a 1.25 inch (31.75mm) square aluminium box section boom and ¼ inch (6mm) aluminium rod elements. All fittings are marine-grade stainless steel. It is shipped part-built with all insulators and the boom-to-mast bracket already in place.

The elements are tightened with an Allen key.

The Stauff element connectors are of a high quality and are secured with an Allen key fastening to hold them to the boom. Fitting the elements to the boom is therefore fast and easy. You need to measure and secure the element end sections (of the loop feed) and fit a feed-line to the antenna. The feed point of the antenna consists of 2 x M4 terminals and the feedline needs to have two terminal connectors soldered to it in order that it may be connected to the beam. Justin has opted for this method over an N-type connector between the coax cable and the antenna on the LFA. He argues that an N-type connector would require 'tails' to be fitted, which would then upset the design of the antenna and affect its designed performance. You should make sure you waterproof it well to prevent water ingress into the coax, using self-amalgamating tape or aerospace-quality rubber sealant.

The antenna has an integral impedance balun, consisting of a quarter-wave stub that connects between the 'hot' (inner core) side of the coax feed point terminal and the boom. This provides a DC ground to the loop arrangement, which further reduces man-made noise. It also acts as a bandpass filter, ensuring very high impedances exist for out-of-band signals. The antenna balun also helps remove eddy currents that can be induced in the boom, ensuring pattern stability and distortion does not occur (as a result of induced current in the boom).

Winding the choke balun.

InnovAntennas also recommends a choke balun at the feed point to prevent the coax from becoming part of the radiating system. Justin says that three turns of RG213 around a former the size of an aerosol polish can is sufficient and he provides instructions on how to make one.

Fine tuning is done by adjusting the loop width for the lowest VSWR. This should be conducted at least one boom length above ground, otherwise false results may occur. If you do not have the means to test and set up the antenna at the suggested height, the antenna can be pointed upwards to the sky at the highest angle possible. This is the method used for the tests. Once set up, it was found that, when raised to its operating position, the antenna exhibited low SWR (below 1.5:1) across the whole band, backing up Justin's claims. It is worth taking a little time to get this setting correct as it is very sensitive and just a few millimetres can make a big difference.

The antenna was mounted on a 40ft Versatower and compared with a competitor's 11-element 2m Yagi. Overall, the results were good and over the test

Close-up of the novel Loop Fed driven element arrangement.

period 2m beacons were heard from all over Europe [from a location in Norfolk – *Ed*] as well as a few contacts being made.

Some tests were made of the antenna's radiation pattern by setting up a transmitter about 1.5 miles from the test antenna. The transmitter was adjusted to give an S9 reading off the back of the InnovAntennas beam. As the beam was swung through 360° a switched attenuator was used to bring the signal level back to S9 and the attenuation details noted every 15°. This was not done under laboratory conditions and multipath reflections and the influence of metal objects and antennas in the near vicinity can obviously have an effect. Nevertheless, the tests were conducted in the real world and went some way to backing up InnovAntenna's claims, derived from the *EZNEC Pro4* modelling, which is shown in **Fig 7.5**.

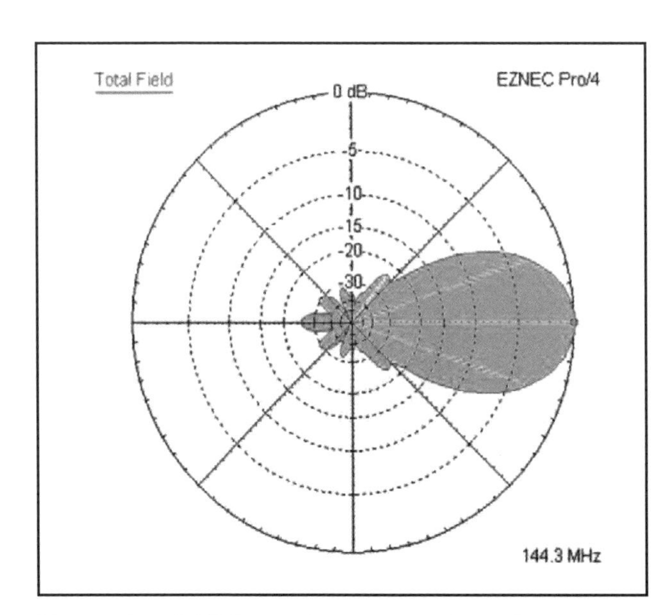

Fig 7.5: The EZNEC plot shows a lack of side lobes.

The tests showed that the antenna's front-to-back ratio appeared to be at least as good as the specifications. What was surprising was the absence of side lobes – an S9 signal was to nulled out completely by turning the beam side-on to the reference signal. This was far better than the comparison antenna could achieve.

It was found that the antenna's rear lobes were much smaller and less distinct than the antenna's competitor. The results showed that the antenna was very broad band, offering a low SWR across the whole of the 2m band. It was also low noise, although we couldn't detect much difference when compared with its competitor. It offers a very clean forward lobe with excellent side and rear rejection, pretty much as predicted by the *EZNEC* plots. No attempt was made to measure the absolute gain figure, although it was found to be slightly down on the 11-element Yagi placed above it (as you would expect). But, as Justin points out: "In the design, gain was not the main objective here and never was. The last few points of a dB that are squeezed from a Yagi antenna destroy it. The antenna maintains high levels of front / back throughout and has no side lobes. This is in the elevation plane in addition to the azimuth plane. As a result (and the use of the antenna balun shorting bar) the antenna has very 'low temperature' – the resulting G/T figure (gain against temperature figure of merit to determine an antenna's ability to receive weak signals) is better than any other antenna of its length.

"The LFA has been cleverly optimised as a complete antenna and offers 'what you see is what you get' results. There are no additional losses to factor in or take off [from] published figures. Also, we quote our F/B and gain figures at one frequency and not the best figure from wherever it appears in the band."

Justin also points out the antenna has a very flat impedance (and therefore SWR) across its operating range and can handle up to 3kW.

The InnovAntennas 9-element 2m LFA Yagi is a well-designed, well-constructed, Yagi that should offer many years of excellent service if built and waterproofed carefully.

INNOVANTENNAS 5-ELEMENT 15M OP-DES YAGI

The previous antenna in this chapter, InnovAntennas' LFA Yagi, used a novel method of feeding a VHF Yagi. For their HF antennas, InnovAntennas have come up with an equally novel feed method: the Opposing Phase Driven Element System, or 'OP-DES'.

OP-DES is patented technology that its designer, Justin Johnson, G0KSC, claims offers maximum performance and a wide bandwidth. Specifically, it offers a 50Ω feedpoint impedance without the need of a gamma match or any other type of

The antenna can be easily carried by one person.

You'll need a step ladder to fit it to the mast.

Make sure you seal the connection to the driven element well.

matching. Instead, the impedance transformation is done within the driven element itself – note the L-shaped ends of the driven element in the photographs.

The 15m Yagi uses many of the construction techniques used in the 2m LFA beam. The main boom is 1.5 inch 16SWG square section aluminium and the elements are lightweight aluminium tubing. This antenna is made with 3/4 inch element sections in the centre of each element, followed by 5/8 inch and 1/2 inch outer elements. The L-shaped end sections on the driven element are 3/8 inch tubing.

Extensive use is made of marine-grade stainless steel for the fittings and the elements are bolted to the boom using InnovAntenna's trademark Stauff connectors with Allen key-driven fastenings.

The 15m 5-element monoband OP-DES Yagi was also reviewed by Steve Nichols, G0KYA, in the August 2012 *RadCom*. The version tested had a 6.849m-long boom, although the design has now been modified with two models being available – one 6.2m long and the other with a longer boom at 8.5m. The turning radius was 4.730m and, fully built, it weighed in at 11.47kg. It could easily be picked up by one person, although fitting it to the tower is another matter. It definitely needs another pair of hands as you are likely to be standing at the top of a step ladder to get it on the mast.

The antenna had a claimed gain at 21.250MHz of 9.36dBi (7.21dBd), with a front-to-back ratio of 26.22dB. In fact, at 15m above the ground InnovAntennas claims the gain is more like 14.56dBi (12.41dBd). Without access to antenna test facilities it is not possible to confirm these figures and so this review really concentrates on the construction and usage of the Yagi and not the accuracy of the maker's figures.

The antenna is best constructed on a large lawn. It comes part-built, with all the insulators and boom-to-mast brackets in place. The individual elements need to be added and the driven element needs to be carefully measured and adjusted once it is fitted. The instructions give you the measurements you need and the whole process is not too onerous.

The end result looks and feels relatively light, but InnovAntennas claims it can withstand winds of 111MPH (a 125MPH version is available upon request). All you need to add is a simple choke balun – InnovAntennas recommends 14 turns of RG-213 or similar, wound on a former such as a large spray can. The ends of the connecting coax need to have two M4 round connectors soldered on for bolting to the boom. It is important that you waterproof this carefully. Take care, use plenty of self-amalgamating tape and silicone sealer, otherwise you will end up with water getting into the braid. The manual recommends using a rubber / aerospace sealant.

The antenna was fixed to a tower at the station of Chris Danby, G0DWV, after setting the height of the stub mast at about 11ft off the ground so that it was possible to slide the antenna on, with the driven element towards the bottom. This worked reasonably well, but you obviously need enough space to do it. The mast was then cranked up a little so that the driven element was about 5 or 6 feet off the ground. At this stage the SWR was tested and found to be lowest at about 21.079MHz. The mast was then cranked up to its full test height of about 45 – 50ft.

The SWR was very low across a wide frequency range: it remained at 1.1:1 across the whole of the 15m band – only rising to 1.3:1 at 20.7MHz and 21.7MHz. Admittedly this was at the end of 120ft of RG-213 coax, but the manufacturer's claims of low SWR and wide bandwidth were substantiated.

In use it was found that the antenna equalled or outperformed a Force 12 C-31XR tribander beam on another tower by about 1 S-point fairly consistently. Noise levels were roughly the same. DX worked included 6V7S (Senegal), VU2XO north of Mumbai, JA3KVT (Japan), V26E (Antigua) and HZ1TT (Saudi Arabia); this during a period of less than perfect ionospheric conditions (SFI: 117, A index: 16, K index: 2).

It was concluded that the InnovAntennas beam works well. At 45ft it performed about as well as the Force 12 C31XR at 65ft. When mounted at the same height, the InnovAntennas beam beat the Force 12 quite consistently. If you were looking for a 15m monoband Yagi, perhaps for a contest station, the InnovAntennas model should be on your list of antennas to consider.

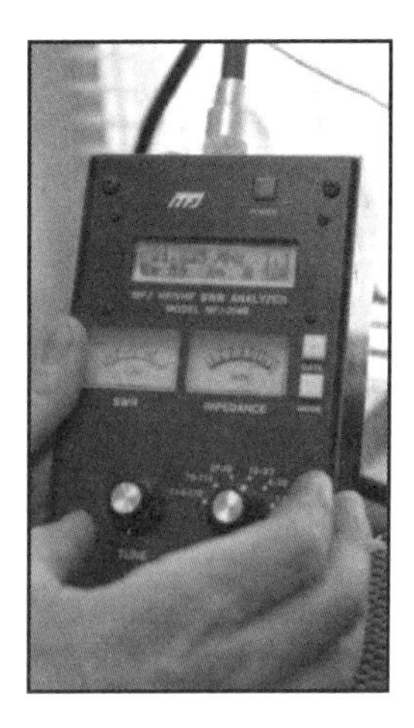

The antenna shows a very low SWR.

INNOVANTENNAS 20/15/10M DESPOLE

The third and final antenna we describe in the trilogy from InnovAntennas also uses the patented Opposing Phase Driven Element System (OP-DES) that its designer Justin Johnson, G0KSC, claims offers maximum performance and wide bandwidth. The InnovAntennas DESpole for 10, 15 and 20m consists of three half-wave dipoles for the 20, 15 and 10m bands. While the 10m element is straight, the 15 and 20m elements are 'kinked', i.e. after going out horizontally for about three metres, the elements drop vertically. The result is a good match to 50Ω coax with the width of the antenna being reduced substantially without having to resort to loading coils or traps. For example, a full-size 20m dipole would be approximately 10.6m wide, but the DESpole manages to fit the band in with a width of just 6.2m.

If the antenna looks familiar that is because it is very similar to the 'Inverted-U' antenna described on page 5 of this book.

As it is the current flowing in an antenna that does the radiating, and the maximum current flow occurs at the feedpoint with zero current at the ends (which are high voltage points), the majority of the radiation comes from the central horizontal section of the antenna. The dropped vertical sections are really there for matching purposes. Of course, on 10m the entire element is horizontal for maximum efficiency.

Another feature of the 3-band design is that only the centre 20m element is actually driven. The other two elements are electrically isolated and work by parasitic coupling (open-sleeve technology).

The antenna is specified as being able to handle 5kW+ and survive 165KPH / 102MPH winds.

The DESpole was reviewed in the December 2013 *RadCom* by Steve Nichols, G0KYA. The antenna arrived in a single long cardboard tube that could be carried by one person. The overall antenna weight is around 7.5kg / 16.5lb. It was assembled at the location of Chris Danby, G0DWV.

The ends of the 15m and 20m elements drop down vertically.

The antenna comes packed in a single long cardboard tube.

The first job is to assemble the centre plates that hold the elements.

The first task was to assemble the centre plates that hold the elements. The two plates bolt together and then the elements are mounted with the Stauff connectors. You then build outwards, adding the elements one by one. These are fixed with a mixture of Allen bolts and stainless steel worm-drive clips. Measure each section carefully to ensure that it is the correct length. If you don't, you'll find the antenna's resonant points may not be where you want them to be.

The 20m element is made with a six section taper starting at 32mm (1.25in) and finishing at 9.525mm (3/8in). The 15m section has five tapers starting at 22.25mm (7/8in) and finishing at 9.525mm (3/8in) while the 10m element has just two sections, the centre being 19.05mm (3/4in) with 15.88mm (5/8in) tips.

You also have to slide two Perspex spacers on to the 20m element, which can be used as guy supports. There is nothing technical about this, but make sure you slide them on at the appropriate point in the build, otherwise you will have to undo all your handiwork later. Leaving the final drooping elements until the antenna is mounted on a mast makes it easier to manhandle the antenna and prevents possible damage to the elements. Once bolted on, using three large U-bolts, the final four drooping sections are added.

You will need to create a choke balun for the feedpoint. This is easy and involves winding about 15 turns of coax around a large spray can or similar former. By laying cable ties on the can before you start you can connect the ties to hold the turns in place. G0KYA and G0DWV wrapped the choke balun with duct tape to keep it together

Only the centre element is actually driven.

Fitting the coax choke balun.

PROBLEMS WITH USING AN ANTENNA ON A BAND IT ISN'T DESIGNED FOR

It is common for amateurs to think they are tuning an antenna with an ATU. When an ATU is connected to an antenna with coax in between, any mismatch is still on the antenna side of the ATU. The ATU merely provides an acceptable impedance for the transceiver to 'see' in order that it will not be damaged. Usually this means that maximum power will leave the antenna too.

The impedance at the output of the ATU will be usually be a long way from 50Ω (unless the rig's ATU is used for its intended purpose, which is to present a more manageable impedance on an antenna for a given band that is perhaps in need of a 'tweak').

And therefore the coax, along with the antenna at the end of the run, are now radiating. Unless an excellent earth system is installed, all manner of issues will result from RF feedback to RFI, but certainly the antenna will not be working as it should or as it is intended to.

Coax losses will be higher too, due to the high SWR on the feedline. For clarity, an ATU should be used:

- To fine-tune an antenna for a given band;
- Where final tweaking is not possible or practical;
- At the end of a balanced feed line;
- At the base of an antenna (long wire or long vertical).

and tie-wrapped it to the antenna feedpoint. The centre conductor and braid of the coax are then terminated in small soldered tags for connection to the 20m driven element. It is important that you waterproof this carefully. The manual recommends using a rubber / aerospace sealant and this appears to be better than trying to use self-amalgamating tape.

First tests showed a 1:1 SWR in the middle of the 15m and 10m bands. On 14MHz the 1:1 point was at around 14.350MHz, measured at the end of 100ft of RG-213 coax. Obviously, the 20m elements were too short, but this was easily remedied with a screwdriver on the worm-drive clips to extend the telescoping sections.

InnovAntennas says that the antenna should offer an SWR below 1.5:1 from 14.00MHz to 14.35MHz, below 1.7:1 from 21.00MHz to 21.45MHz and below 1.9:1 from 28.00MHz to 28.60MHz and our figures backed this up once it was tuned correctly. The antenna worked very well: it is essentially a unity gain antenna, so don't expect it to perform like a Yagi. The InnovAntennas dipole was unsurprisingly usually about 2 – 4 S-points down on G0DWV's Force 12 beam at about 65ft but often there was little difference in signal strengths between the two antennas. Stations from Hong Kong, India, Canada (Niagara Falls) and Senegal were worked or heard, with 15m being the 'money band' at the time of operating.

G0KYA and G0DWV even managed to get the transceiver's internal ATU to tune it on 17m with reasonable performance. It is a 'get you going' antenna on the 17m and 12m bands, though don't expect all transceivers' internal ATUs to be able to match the antenna on these frequencies: the 'raw' SWR on 17m was 3.5:1 and on 12m it was 5:1 at the end of the coax. While stations were worked on 17m (including K6YRA) signals on 12m were down about 3 – 4 S-points on a half-wave sloper at about 55ft (please see the sidebar).

Overall then, the 20/15/10 DESpole rotating dipole offers three-band performance in a reasonably

The DESpole needs to be mounted on a fairly sturdy mast or pole.

lightweight package. Reasonably lightweight as you need a fairly sturdy mast or aluminium pole to support the weight – this isn't for mounting on a one-inch pole or fibreglass fishing rod. The maximum turning circle is only 6.2m, which means it is more likely to fit into narrow spaces.

InnovAntennas also produce versions that cover 15, 10 and 6m; 20, 17 and 12m and 30, 20 and 15m. This means that you could mount the 20, 17 and 12m variant underneath a triband Yagi, giving you access to the 17 and 12m bands and perhaps bi-directional performance on 20m, which can often be helpful in contests and the like.

Further information on the DESpole can be found on the InnovAntennas website at www.innovantennas.com

THE WHIZZ WHIP

This 'miracle' antenna design seems to have existed in many guises and the Whizz Whip is the latest incarnation. (A similar home-made antenna was designed by John Goody, M1IOS, and is featured in Chapter 2 of this book.) This 'Whizz Whip' version is made by Moonraker (UK) Ltd (www.moonraker.eu) and it claims to be a universal QRP (10W) transmit and receive antenna for 80m through to 70cm with extended receive coverage of 600kHz to 500MHz. It was reviewed in the December 2014 *RadCom* by Mike Richards, G4WNC.

There are two parts to the Whizz Whip, a 130cm telescopic whip antenna and the main control unit. The included instructions comprised a single A4 sheet with the specifications and a few operational instructions.

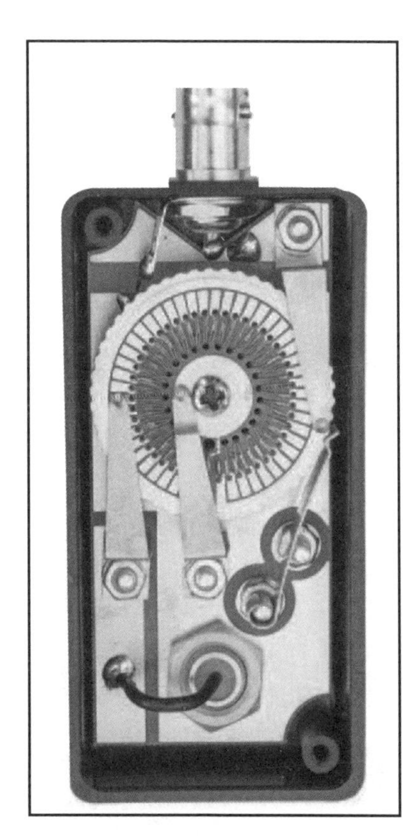

Internal view of the Whizz Whip antenna.

The Whizz Whip has a BNC connector at the top for connection to the telescopic whip and a PL-259 plug on the front panel for connection to your rig. The normal configuration would be to screw the PL-259 connector directly into the antenna socket on your rig. This works very neatly for small portable rigs like the FT-817 but didn't fit so well on the FT-897D. Ideally, you need a rig where the antenna socket is mounted towards the top of the rear panel so you can access the rotary tuning knob on the front of the Whizz Whip. In its simplest form, this is all you need to get operational.

In cases where the Whizz Whip doesn't fit neatly on the back of the rig you can use a coax extension cable. However, there's no mounting system for the control unit so it becomes a bit cumbersome when attached to the whip. One solution could be to mount an SO-239 socket on a vertical metal sheet and attach the sheet to a firm base for stability. That way the Whizz Whip could plug into the new socket where it could then be wired back to the rig. You can also use the Whizz Whip control unit as a universal QRP matching unit to all manner of wire and whip antennas.

The photo shows that the main component in the Whizz Whip is a multi-tapped inductor combined with a rotary switch. The switch has 48 steps and provides access to 47 different inductor taps. This seems to be an L-match arrangement but with a much larger number of taps than usual. The large number of taps is the key to the Whizz Whip being able to work over such a wide frequency range. For the VHF / UHF coverage, matching is achieved simply by trimming the length of the telescopic whip element.

Before starting to operate with the Whizz Whip, G4WNC made some SWR measurements using a YouKits FG01 antenna analyser. The first set of measurements were taken with only the telescopic whip attached and the Whizz Whip's tuning control adjusted for the lowest SWR. In this configuration, the Whizz Whip is very sensitive to objects in the immediate proximity as the case is made from ABS or similar so there is no screening. To get reliable measurements the antenna was secured by holding the coax connector in a clamp. For the second set of measurements a 15m counterpoise was attached to the screen of the coax connector and ran out along the ground. As you can see from **Table 7.2**, it was possible to obtain a workable match on most HF bands. You can also see that the counterpoise improved the performance on many bands. Obtaining a good match on 3.5MHz was extremely difficult as everything was 'hot' and just changing the operator's position affected the result. However, with more time and some tinkering with the counterpoise length the match on 21MHz could have been improved.

Band	SWR with Whip only	SWR with Counterpoise
3.5MHz	3.6	3.2
7MHz	1.8	1.6
10MHz	1.4	1.4
14MHz	1.6	1.2
18MHz	1.3	1.3
21MHz	2.1	2.3
24MHz	1.7	1.1

Table 7.2: Whizz Whip SWR results.

For the on air tests G4WNC was set up in the garden, initially just using the telescopic whip. The bands were fairly quiet so PSK31 was used as there's almost always activity on 7MHz, 14MHz or 21MHz. The first problem encountered was lots of RF getting into the laptop computer, despite running just 5 watts of RF. To help tame things a 15m counterpoise was attached to the ground connection on the rig and ran out across the garden. This improved the situation but there was still some RF getting into the PC so some clip-on ferrites were attached to the audio leads between the rig and the laptop. That did the trick. With this arrangement, a few European stations were worked on PSK31. The best distance on the day was northern Italy, which is not bad for such a simple set up on a flat band. 599 reports were being received, but that's a bit meaningless as everyone seems to give 599 on PSK! What was more telling was that G4WNC was getting immediate responses to his calls: a sure sign of a decent signal.

Although many will be sceptical of the claims of these 'wonder' antennas, this one works, providing the operator takes some care. For HF use it really needs a counterpoise but that is generally easier to set up than a suspended wire antenna. For VHF / UHF operation you simply extend or retract the telescopic whip to achieve a good match. Before you buy, you also need to consider how the Whizz Whip will attach to your rig, SWR meter, etc, as it is not self-supporting. As the review was completed, Moonraker reported that they are currently looking at different mounting options and hope to have some new accessories available for the Whizz Whip.

OPTIBEAM OBW10-5 MULTI-BAND BEAM

The Moxon Rectangle is now a well-established design, as is the Yagi of course, but the OptiBeam OBW10-5 combines Yagi and Moxon with a five-element driver cell all together in a single novel antenna. It is the German manufacturer's answer

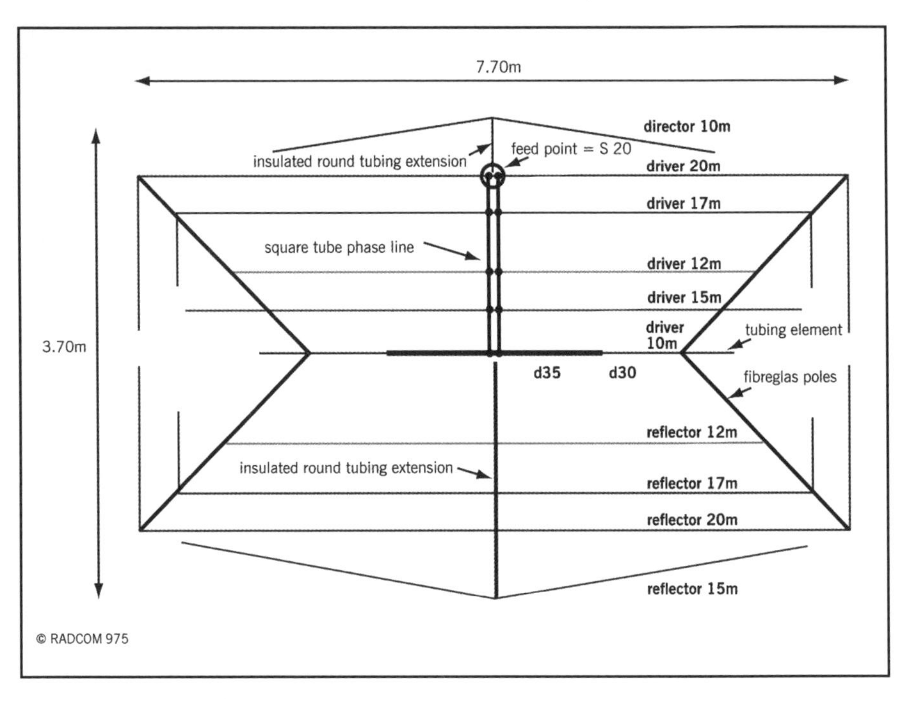

Fig 7.6: OptiBeam OBW10-5 layout.

to providing a reduced size, lightweight beam covering all five HF bands from 14 to 28MHz with very little in the way of compromise as regards performance. The antenna was launched in 2006 and is still in OptiBeam's portfolio (www.OptiBeam.de) as of March 2015, so this highly innovative design has certainly stood the test of time.

The OptiBeam OBW10-5 is a compact five band wire antenna with two elements on each band. The four longest elements, those for 14 and 18MHz, are Moxon rectangles, four more elements are conventional straight Yagi elements, while the remaining two are also Yagi-type elements, but bent into a shallow 'V' shape in order to save more space. **Fig 7.6** shows the layout.

The main structural component of the antenna is the five-element driver cell using a rigid phasing line which results in a very close match to 50Ω over all of the HF bands. A 1:1 balun capable of handling 3kW is available as an option. Nine out of the ten elements are made of very strong and flexible stainless steel wire, dramatically reducing the weight, wind load and – most importantly these days – the visual impact of the antenna. Although stainless steel has much higher bulk resistivity than aluminium, according to modelling with *EZNEC* any losses caused by this are very small, maybe just a fraction of a dB.

The antenna consists of a frame made up from the boom, half of which is the phasing line, and the 28MHz element. This is extended with four fibreglass extensions as a pair of Vs (see **Fig 7.6**). The wire elements are attached with non-conducting blocks to the frame. The elements are tensioned using elastic cables and by stretching the frame itself. Most cable to element connections use an insulator and a stainless steel shackle.

All the nuts and bolts are stainless steel, and the nuts are self-locking. OptiBeam supplies a small set of tools, a tube spanner and Allen keys, which facilitate assembly, and a few spare nuts and bolts – just in case! All of the various sub-components are very clearly marked and fitted exactly as described in the manual. Some care in assembly is necessary so that the elements are evenly tensioned. A truss is used to ensure that the antenna does not 'flop' down when it

is on the tower. This means that there must be about a 1.5m stub mast above the antenna.

The OBW10-5 was reviewed by Dr Bob Whelan, G3PJT, in the March 2007 *RadCom*. Although this antenna can be assembled by one person, the final tensioning of the assembled frame and elements is much easier if two people can each pull from either end and tighten up the frame when the elements are taut.

Getting the OBW10-5 on to a tower or mast requires some forethought. At 14kg the antenna is reasonably light compared with many all-aluminium Yagis and can easily be picked up in one hand. However, as it has closed element ends (the folded-in ends of the Moxon elements), it has to be lifted over the top of the stub mast and cannot be fed in from the side as one can do with a normal Yagi. This isn't a problem in itself, unless for example you have other antennas such as a VHF beam on the stub mast.

Using the OptiBeam dimensions to tension the truss it was correct first time and when the tower was wound vertical, the antenna looked slightly upward bowed, just perfect.

G3PJT wrote: "When the OBW10-5 was up in the air I was immediately stuck by how small and compact it looked... It is a *lot* less visible from a distance [than G3PJT's other OptiBeam]... It is also light and well balanced, so a small rotator would handle it easily and there are no worries about wind resistance either. It sailed through the storms at the end of November [2006] without any problem."

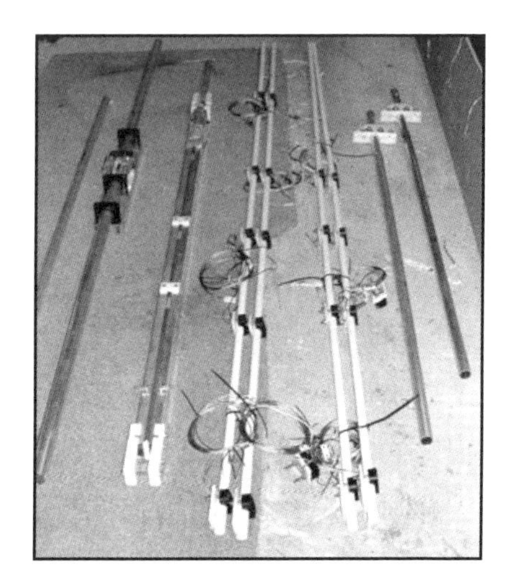

All the bits after unpacking.

Bands covered		20 / 17 / 15 / 12 / 10
Gain (dBd)		4.2 / 4.4 / 4.2 / 4.4 / 4.8
Front to back ratio (dB)		21 / 17 / 15 / 15 / 20
SWR:	14.00 - 14.13 - 14.35	1.8 - 1.3 - 1.8
	18.07 - 18.14 - 18.17	1.5 - 1.4 - 1.5
	21.00 - 21.25 - 21.45	1.5 - 1.1 - 1.5
	24.89 - 24.94 - 24.99	1.4 - 1.3 - 1.4
	28.00 - 28.50 - 29.00	1.5 - 1.2 - 1.6
Elements		10 (9 wire elements)
Active elements per band		2 / 2 / 2 / 2+ / 2+
Longest element (m)		7,70
Boom length (m)		3,75
Turning radius (m)		4,30
Feedlines		1 Coax 50Ω
Weight (kg)		14
Wind load at 130 km/h		242 N / 0.30 m

Table 7.3: OptiBeam OBW10-5 manufacturer's specifications.

On all bands the antenna behaves as a two-element Yagi. On 24 and 28MHz it is claimed that there is some extra gain from interaction with unused elements. While it was not possible to measure forward gain, a series of measurements of the front-to-back ratio under various conditions showed that it was very close to the claimed specifications (**Table 7.3**) and under some conditions exceeded them. The practical SWR measurements with the antenna at 18m and 20m of feeder were substantially to specification as well.

Looking at the performance differences between various antenna designs shows that the OBW10-5 gives up less than a dB to a beam with conventional elements like the OptiBeam OB9-5, which weighs in at 25kg and has a boom length of 5.1m against the OBW10-5's 3.7m! Indeed the OBW10-5 is only about a dB below the gain often quoted for a typical 3-element Yagi. Of course with a lightweight antenna one can often mount it higher than a heavier one and thus exploit height gain for even better performance, and be able to leave it up during the winter gales.

G3PJT summarised, "The OBW10-5 seems to me to be an excellent antenna. Frankly its performance is very good for its size. It should easily outperform many so-called miniature beams especially if they use lumped loading devices or traps. The OBW10-5 should be considered by anyone who needs a low visual impact antenna, which at the same time is lightweight and well constructed. It should last for years."

Chapter 8
PICaYAGI

THIS CHAPTER IS devoted entirely to a single antenna. It is a major constructional project – the PICaYAGI, designed, built and described by Peter Rhodes, G3XJP. It was published in *RadCom* in four parts, from July to October 2011, and is printed here in its entirety and as a single feature for the first time.

PART 1 – INTRODUCTION

In 1926, Yagi and Uda invented their beam antenna. We were soon building monoband Yagis for 20 / 15 / 10m and, shortly thereafter, the popular trapped 3-element tribander. Then along came the WARC bands and since then we have been struggling for a viable single beam to cover all five bands. Several different commercial approaches have evolved over the years but none of them met my requirements – for reasons discussed shortly.

PROJECT SUMMARY

This is a constructional project to build your own 5-band (20m – 10m) Yagi that automatically grows and shrinks its element lengths as a function of your transmit frequency. This is intended as a repeatable design but, by the very nature of mechanical construction, I would expect significant variations in implementation. This is not therefore a prescriptive approach. You may need to choose different materials (e.g. tube sizes) and you may think of better engineering implementations. To this end, some of the blind alleys and false trails are also mentioned since there is rather more learning to be derived from mistakes than from success.

All my previous PICaPROJECTS **[1]** were rigorously tested for reproducibility by multiple builds prior to publication. This one is different. In fact, as of Spring 2011, my PICaYAGI is the only one in existence. So although the formal engineering drawings have been rigorously checked for accuracy, this can never be a substitute for verification by building. Further, this design uses some novel concepts that inevitably increase the risk. So, although mine works brilliantly, there can be no absolute guarantee that this is reproducible. There is much good faith and conviction – but no warranty.

This project also has attitude! This is homebrew amateur radio so the project objective is about more than just acquiring a beam. It is also about increasing personal skills and learning through doing. It has been an incredible source of learning for me and was chosen for that very reason – and also because I wanted a decent beam to complete my homebrew station.

Photo 1: G3XJP/M during early proof of concept trials in Spring 2010.

YOUR INFRASTRUCTURE REQUIREMENTS

As in all the other PICaPROJECTS, the need for any serious workshop technology has been carefully avoided. There is no welding, lathe or milling work. There are some GRP (glass reinforced plastic, or fibreglass) parts to be fabricated, but nothing so serious that they can't be addressed with a typical car repair kit. You will need an old PC and the means to program a PIC. The software itself is provided as both source and object code at no charge – strictly for your personal amateur use. You need a cheap pop rivet gun and the ability to shrink heatshrink tube. Some RF performance modelling software is useful – and the best, in my experience, is free. You also need to be able to make your own PCB – see **[1]**. This antenna can be built entirely from your own amateur resources and you don't need to pay anyone else to do any of it for you. So I won't be authorising any commercial PCBs to circumvent the project objective, though there may possibly be scope for bulk buys of raw materials.

ACKNOWLEDGEMENTS

This concept has been evolving since November 2008 with the help of a large number of people – a few of whom have been active and completely indispensable. My special thanks to Dave, G3SUL; Harold, W4ZCB, and Chris Stake. And not least Fran, who has been frequently called upon for a second pair of eyes and hands – and for her ability to proofread stuff she doesn't claim to understand to see if it makes sense.

The latest software and the latest updates are available at **[1].**

MY PERSONAL REQUIREMENTS

Like everyone else I want multi-band coverage but, being greedy, I also want mono-band performance. Actually, I want better. I want mono-*frequency* performance. That is, a beam tuned for the spot frequency I'm on at the time, not compromised by the need for coverage of an entire band.

But for me, the dominant consideration has to be visual impact and, at least in the UK, I suspect it is for many others too. Frankly, I don't want some monster polluting my skyline and I can't imagine why any of my neighbours would either. So I guess my fictional engineering figure of merit would be along the lines of 'dB performance per metre of obstructed sky'.

HOW DO OTHERS DO IT?

Always a good first question! Taking the 3-ele Yagi as a performance datum, there are several radically different strategies for 5-band beams on offer. One approach is the interlaced 5-band 2-ele Quad. Then there are log periodic arrays and log-Yagis. Various designs interlace multiple Yagis on the same boom, some also with log cells. A radically different approach is taken by SteppIR™, whose Yagi elements change electrical length as a function of frequency. None of these approaches satisfy my basic requirements. Why not?

- A Quad with spreaders long enough to cover the 20m band is a highly visible 3-dimensional structure
- Any log periodic array gives relatively poor performance per element and so uses its quota of 'obstructed sky' inefficiently
- On that same scale, 5-band interlaced designs are even worse since most of the elements do nothing most of the time – so, when you hear someone on 20m proudly announce that he is using an 11-el Yagi, I just hope he realises that only 3 of them are actually in use
- The SteppIR Yagi is fixed in mechanical size (as opposed to electrical size) so whatever band you are on – and even when not in use – the visual impact is always worst case, namely that of a 20m beam.

For me, all these approaches deliver ~3-el Yagi performance but score poorly on visual impact. With commercial offerings, there are also significant value for money considerations and some performance issues. A glance at any price list shows the former is self-evidently an issue. The latter will be addressed later.

DREAM ON

For several years I have been idly contemplating ways of building an HF Yagi that grows and shrinks in size as a function of frequency. Plasma? Conductive liquids? A scaled-up version of those evil party blow-out toys? In September 2008, with the prospect of some sun spots, I decided to get full-time serious. And with the end of 10 years of previous project support commitments, I was able to.

I want something about the size of a 10m Yagi when not in use (i.e. the vast majority of every 24 hour day). There are three themes that were pursued in parallel and that needed to converge to turn the dream into reality.

Changing element resonant frequency. I began by looking at the most obvious approach, namely telescoping aluminium tubing. The fundamental issue is that any rubbing aluminium surfaces will promptly seize. Any other metal is way too heavy and any metal-to-metal contact has a very short life outdoors. After much debate Dave, G3SUL, came up with the brilliant idea of using heatshrink tubing applied to an inner sliding aluminium tube as a dielectric layer. This provides a reactive coupling between the sliding and fixed tubes. The engineering detail follows later – but this has remained the approach ever since.

Frequency span is, however, a problem. The essential issue is that from the bottom end of 20m to the top end of 10m is more than a 2:1 ratio. And the nature of any telescoping tube arrangement (given an overlap) is that the length range is less than a 2:1 ratio. The reality is worse than that because the smaller diameter inner tube increases the resonant frequency, as does any element droop. Dielectric coupling also reduces the effective length near full extension. So for many months we looked at telescopes within telescopes and some extraordinary pulley arrangements to drive it all in and out. None of it felt engineeringly realistic.

At about this time I decided on a Yagi with a fixed length boom and fixed element spacing – simply to bound the task. **Fig 8.1** shows the chosen tubing configuration. The essential approach was to make the fixed inboard length of each element from two ~4m tubes, overlapped, clamped together and fixed to the boom. This gives a net length of just over 2m per side, which can be pre-adjusted by altering the overlap to make a director, reflector or a driven element for the top end of 10m. Then, with sliding inner tubes also ~4m in length and with a minimum overlap of ~1m, these elements can be extended to over 5m per side to hopefully cover down to the bottom end of 20m.

Motive power. I started out looking at pneumatics. This has the appeal that air is invisible, not very heavy and comes for free. I was already using compressed air for the mast, though that is not a prerequisite for others and certainly not part of this project. Ultimately I think I could have made it work but the problem lies in

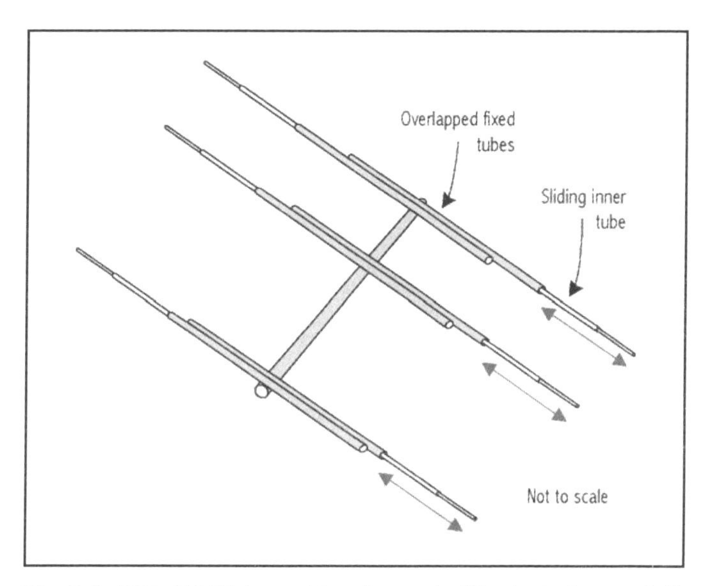

Fig 8.1: PICaYAGI tuneable elements illustrated near full extension.

Photo 2: First installation on the pump-up mast in late Spring 2010. The mast also helps reduce visual impact – the castle on the hill shows why I need all the help I can get. I also have a bit of a ground slope challenge.

measuring the resultant element length. There is little virtue in formulae like '2.3 bar applied for 3.7 seconds = QSY down 50kHz'.

Stepper motors are an instinctive choice for repeatable positioning but, given the torque requirement, they were immediately ruled out as too bulky, too heavy and too expensive. On the basis of size, weight, torque, price and availability I finally settled on a £6 electric screwdriver from a DIY chain. This has a 3.6V motor and an epicyclic gearbox that delivers 1.5Nm of torque. I use one per element.

Measuring element length. My first approach was using sonar range-finding with a piezo sounder on each element tip and an electret microphone near the centre of the boom. Unfortunately, the speed of sound in air varies significantly with temperature. The complication of adding a temperature sensor and all the inevitable calibration curves in the software felt entirely disproportionate. So I pragmatically abandoned this otherwise excellent bird scarer.

Laser range-finding is prohibitively expensive.

The obvious approach was to attach some sort of 'string' to the element, spool it on a shaft and then count turns with a shaft encoder. Having made that leap, the obvious next step was to use that same string to also pull the element in and out – abandoning the pneumatics.

For the string itself, my first thoughts focussed on beaded cord, as commonly used on domestic window blinds. This is basically nylon braid with moulded nylon balls at about 10mm intervals. The idea was to drive it with a sprocket, much like a water wheel. The great virtue of this approach is that you can pass the cord once round the sprocket and the same length of cord is used for both IN and OUT purposes. In other words, there is no spooling requirement. In trials, I could not get it to work effectively because the diameter of the sprocket needed to grip several balls was too large; the screwdriver didn't deliver enough torque. Also, fitting the ball diameter in the gap between the fixed and moving parts of the element was distinctly marginal.

On the recommendation of Paul, G0ILO, I went instead for braided fishing line. I tried several on the basis of their advertising claims but 'zero stretch' proved to be somewhat overstated. Some of them appeared to just go on stretching forever! On advice from Steve, G4ZBV, I avoided anything based on Kevlar on the grounds that it is self-abrading, i.e. fine until you spool it tightly.

Finally, I got in touch with some serious kite-flyers on the West Coast and ended up with some Dyneema braided braid that is less than 0.5mm diameter and has a breaking strain of 80lb / 36kg. That is incredible performance. It is also essentially (if not actually) zero stretch at the loads I am applying, which greatly simplifies the inevitable software task of converting counted shaft encoder slots into a repeatable element length. This wonder braid is inexpensive and readily available around the world.

SUMMARY SO FAR

Having settled on the underpinning mechanical technology, there followed the engineering to make it work reliably and all the issues of RF performance. There were still several outstanding conceptual problems:

• Would this approach produce enough change in element length? Indeed, how much is needed to achieve the frequency span?

• What sort of boom length would be needed to give good performance on all 5 bands? And what exactly is "good performance"?
• How do you feed a driven element – that cannot be split – across all 5 bands?
• Could it all be made to withstand a wind?

Time for some antenna modelling.

WHY MODEL AT ALL?

For the simple reason that it would be essentially impossible to find the optimum dimensions of the real antenna by cut-and-try. In this design, you start out with at least five interdependent variables: element lengths, boom length and element spacings, to say nothing of the feed arrangements. The task is to determine the best dimensions such that it will perform well over all five bands and at the same time, have enough mechanical strength to have a reasonable chance of survival.

The task is iterative and it would cost a fortune in time and materials to attempt on a real antenna. Frankly, you would be unlikely to ever get there. Most likely you would probably produce something that worked, but was far from optimum. Worst of all, you would never know it. This topic is therefore particularly relevant if you are considering changing any of the element spacings or tube diameters – and of interest to anyone designing or assessing any antenna.

The subject of modelling splits, by tradition, into the inter-dependent tasks of ensuring mechanical integrity and optimising RF performance.

MODELLING MECHANICAL BEHAVIOUR

How do the professional designers do it? A commercial manufacturer cannot know, for example, the likely wind speeds or ice loading at the customer site because, obviously, they are all different. So they have to design to some established standards, which typically have generous margins for error. These are typically national standards so there are then all the issues of certification if you want to market the product abroad. All this costs much money and it inevitably finds its way to the bottom line price.

For the one-off amateur it can (and should) be very different. Round here, for example, earthquakes, tornados, hurricanes and significant ice storms are all very rare events. Not unknown, but we are talking typically once in 50 years stuff here in up-state Herefordshire.

Would you really want to design to handle them all? Because if so, you go very fast down the road of diminishing returns and end up using gargantuan tube wall thicknesses and need a beefy (but very low) tower to hold it all up – and a rotator that would start a jumbo jet on a cold morning.

A lot of these rugged designs originate in the USA, where they certainly do need to design for significant environmental extremes. By UK standards they have some extremely tolerant planning legislation for amateur antenna installations. Pragmatically, the most likely outcome of designing to handle those same risk levels in most parts of the UK is that you would end up with no antenna in the sky at all.

Ultimately, only you can assess the risks and the consequences that you face. The realistic consideration I believe is the price of failure. I'm lucky (actually it's long term life-style choices) to have my mast in a field that is well out of range of any human activity. The only person that ever ventures into the drop radius of the beam is me – and I can choose when I do so. But obviously, I still want to design for reasonable survivability of my precious antenna.

Needless to say, we are always trying to hold down the total weight, turning radius and crosssectional area in order to minimise cost, visual impact, wind loading, mast requirements and rotator torque. However, all this down-sizing tends to work against both RF performance and the ability to withstand ice loading. It is a vicious circle. I don't have a wind-tunnel here and I'm also not inclined to add weights to the elements to discover when they would break under ice loading. My pragmatic approach in the end was to design for RF performance using materials that were available and 'felt' as though they might handle the stresses – using the *ARRL Antenna Book* HF Yagi mechanical designs for guidance.

About the same time Dave, G3SUL, pointed me in the direction of the basic equations for cantilevered beams. This is frightening stuff. Each half of an element is a cantilevered beam fixed at one end. The equations show that the droop of an element under its own weight rises as the cube of the length. So if the tube diameter or the wall thickness is inadequate, you can easily get to the point where as you extend the element, the tip is moving rather more down than out. It is how to make an inverted-U. Anything even approaching this is a disaster for achieving the frequency span target, let alone mechanical integrity. It is why the element diameter is tapered down as we approach the tip. We also need to find the optimum wall thickness, since increasing it is a trade-off between adding strength but also increasing weight. After that, any ice loading just increases the weight and windage.

At this stage I made the decision that I was only ultimately concerned with stresses on a *parked* PICaYAGI, with the elements fully retracted. For visual impact reasons, I already have the discipline of parking the elements whenever it is not in use – and certainly overnight. And if we have gales or any evidence of serious icing then I simply forgo the pleasures of the lower frequency bands.

The practical approach I use is to wind the elements out and, if they start flaying about, I wind them back in and move up a band until they are not. You quickly learn to correlate the movement of tree branches with what is reasonable. You may consider this approach to be very amateur and I would proudly agree. It is no more onerous than asking a sailor not to raise all the sail in strong winds and, if gale force, to stay in the harbour. It happens very rarely.

The upshot of all this is that the parked elements are comparable, dimensionally, with the design of the 10m Yagi in the *ARRL Antenna Book*, falling between their heavy duty and medium duty designs. The medium duty design will handle "wind speeds of 96 mph with no icing and 68 mph wind with ¼in of radial ice." That will do for me!

To balance the rotator wind torque, I also need to add a small torque compensator plate to the boom.

Finally, in order to reduce friction on the sliding tube, it is important that the fixed tube remains as straight as possible as the elements extend. To this end some vertical cord bracing is essential. However this bracing has no other function when the element is parked than to withstand any ice loading. So that is indeed the icing on the cake!

RF PERFORMANCE MODELLING

First we need to define some terms. I both regret and resent having to do this here – but 85 years since Yagi and Uda (and despite heroic efforts by the ARRL), there appears to be little consistent practice out there. The three parameters we typically want to tune for are gain, front-to-rear ratio (F/R) and SWR. Guess what? These – and the definition of the environment – are all open to abuse.

Environment. We can define our performance in hypothetical free space, or at a specified height over specified ground. The merit of free space is that nobody has access to it so we are all equal. Thus we can validly compare one configuration to another or one antenna to another on the basis of calculated performance. The merit of using real ground is that you can make real measurements and it is generally possible to determine if you have improved your antenna – but it is desperately difficult to define the measurement circumstances such that you can validly compare it with others.

All PICaYAGI performance numbers used to evaluate concept feasibility were calculated in free space. All subsequent measurements of achieved performance were made under some precise circumstances that I will carefully define later.

Gain. This is the performance in the main lobe relative to either a dipole in free space (dBd) or relative to an isotropic point source in free space (dBi). Either will do fine and one can convert using the formula dBd + 2.15 = dBi. The confusion creeps in when one unit is maliciously used in free space and the other over real ground. You end up comparing apples with oranges.

Back and Rear. *Wikipedia* (and others) say front-to-back and front-to-rear ratio are the same thing. Not here they are not. I use the definition in the *ARRL Antenna Book*. Front-to-back ratio (F/B) is the gain ratio between the main lobe and the lobe at exactly 180° off the back. By contrast, front-to-rear ratio (F/R) is the gain ratio between the main lobe and the worst lobe anywhere to the rear. Both are measured in the same plane as the gain. F/R is the more strenuous criterion and is always the one used here.

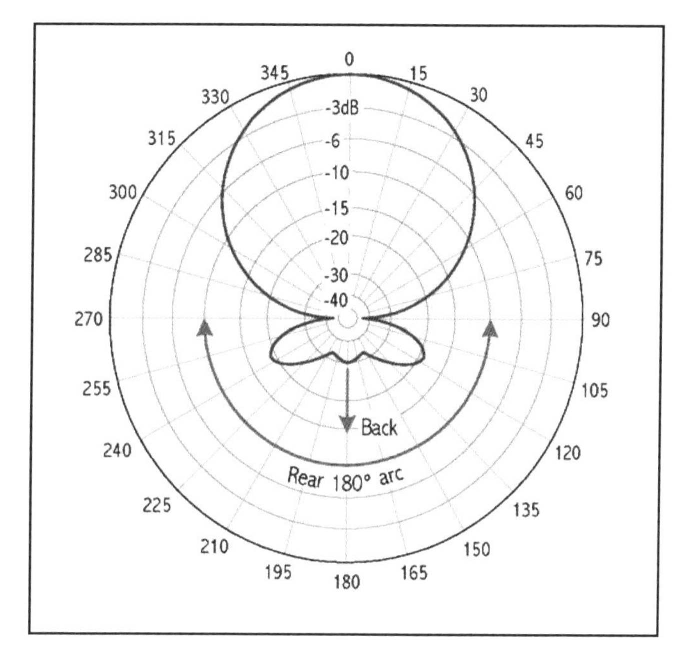

Fig 8.2: Performance plot of a 3-element Yagi tuned for best F/B – at the expense of F/R.

A plot for a typical 3-element Yagi is shown in **Fig 8.2**. This antenna has been tuned for good F/B and it achieves a very attractive sounding 30dB. This tremendous performance directly backwards is always accompanied by unacceptable rear lobes in other directions. The F/R is a much less exciting 18dB. Except for specialised applications, quoting only F/B is misleading and unhelpful, not least because QRM rarely chooses to be exactly on a reciprocal bearing.

SWR. The other modelled performance parameter that must never be taken for granted is SWR. One behaviour that modelling any antenna teaches you is that if you allow the SWR to drift up even slightly then it is possible to produce significantly better apparent numbers for the gain and rear performance.

Provided that the beam can indeed be tuned for unity SWR at the chosen feed impedance then I don't accept anything over 1.01:1 as a mandatory maximum. If it's more than that I just keep on optimising until it isn't – while typically watching the gain and rear performance fall way. That way the results are validly comparable with others. In real life, I'm perfectly happy with modest SWR on the feeder but in trying to optimise a design, the practice of letting the SWR drift up is one best confined to the production of doubtful advertising copy.

IS MODELLING EASY?

In his penultimate 'In Practice' column **[2]**, Ian White, GM3SEK, said that one (of many) aspects of an antenna design that we need to consider is "ease of computer modelling and whether a design can be converted into real-life hardware with a minimum of uncertainty." Although Ian was referring at the time to VHF / UHF long Yagis, the sentiment is equally true for any Yagi.

However, at VHF / UHF, the task is more straightforward – because you mostly use straight, untapered elements. There tends to be a lot more of them, of course, but that only increases the scale of the modelling task, not its validity. These elements don't fall into the grey zone of what the modelling software will validly handle. This got me thinking. Could it be that most of the popular HF Yagi designs are what they are only because they could be easily modelled? Is that why they are so unadventurous?

ARRL does not publish antenna advertisements that include gain figures unless the advertiser can provide certified results from an antenna range or antenna modelling data that indicates that the antenna *should* have the gain described **[3]**. Suitable modelling packages include *YO* (*Yagi Optimizer*, for Yagis only) or the latest version of *NEC*. One can only applaud their desire to eliminate misleading advertising claims.

PICaYAGI can be validly modelled with either of these packages – but it is not easy. Why? Because literally every unusual aspect of this design falls outside the scope of what these packages will handle without significant creativity.

I confine this discussion to the use of *NEC2* – since that is what I actually used. It is free and later versions (which are not) offer little if any incremental benefit for Yagi design. The trap for the unwary is that *NEC* itself will not prevent you from entering pretty well any complex structure you like. All it needs is the radius of all the tubes and the co-ordinates of both ends. You can add reactances and a feed point and other forms of discontinuity. Having entered all those physical parameters, *NEC* will then do the sums and give you the answer. This is great, but in the case of PICaYAGI (or anything else with unusual features) you would be well-advised not to believe it.

This is not a criticism of *NEC* in any way. It is great – if used carefully and within its specified limits. All mathematical models depart from reality in some ways and the *NEC* documentation goes to a lot of trouble to point them out.

The problem is that if you model PICaYAGI literally, you break most of the rules, produce some spectacular numbers and, in many cases, no warnings. So, regrettably, it doesn't meet GM3SEK's criterion for ease of modelling.

It is rather difficult – and this is why:

Element cross-section. As shown in **Fig 8.1**, the centre of the elements has an overlapped section. This is simply not allowed and you have to replace those two clamped tubes with one round tube of somewhat larger diameter. How much larger?

Tapered elements. NEC is notoriously inaccurate in modelling any discontinuities in element diameter. PICaYAGI has a large % step in diameter where the sliding tube enters the fixed tube. In general, there is an excellent way to handle this, called Leeson's Correction. It replaces all the tapered tube lengths with one of the same total length but with one equivalent diameter. But Leeson's Correction has several restrictions and infringing any one of them prevents us from using it. Needless to say, PICaYAGI infringes the whole lot. Specifically, you are not allowed any reactive loads on the element and the elements must be perfectly straight. No droop!

Photo 3: PICaYAGI at full element extension showing modest element droop.

Element droop. This causes a problem in its own right since, if you model it as a series of wires at increasing angles, the model rapidly becomes unstable – producing wild swings in output for small changes of droop.

Feed point. My tentative idea for the feed arrangement has a fine and detailed structure compared to the elements. It uses differing diameter tubing at apparently arbitrary angles to the driven element. There are precautions you can take with careful segment alignment – but you are taking a big chance.

So how to get around all this? The only way I know is to take the real element, put it up in the sky, measure its self-resonant frequency and then enter it back into NEC as tubing of the same length but with one net diameter that resonates on the same frequency. This is in effect manual Leeson's Correction but without the restrictions. **Photo 4** shows some early measurements being made on my driven element. These were later refined at the intended install height. You will have spotted the catch. You have to commit to the materials to build the antenna to build the model to prove the antenna will work. In reality, you creep tentatively up to that point with an ever increasing sense of confidence.

Photo 4: Scoping the frequency span of the driven element.

Finally, having built the model you can (and must) run validity checks to test whether you have got adequate segmentation and that the sum of the power radiated in all directions is equal to the input power. These tests give you significant confidence that the modelled performance results can be trusted.

CHOOSING A MODELLING PACKAGE

There are many PC packages around which provide a user interface to the ubiquitous *NEC* core – with much variation in price and facilities. I am indebted to Ray, WB6TPU, for pointing me towards *4NEC2* by Arie Voors **[4]**. It is an absolute delight to use. Arie has devoted countless years to getting this package right and he makes it freely available, in the true spirit of amateur radio.

The critical *4NEC2* feature from this project's perspective is the inclusion of a genetic optimiser. Most traditional antenna optimisers use a hill climbing approach to optimise your specified variables (e.g. element lengths) against your chosen criteria (e.g. Gain, F/R, SWR). They will indeed get you to the top of the hill. But what they don't tell you (simply because they don't know) is that it may not be the highest hill in the mountain range. By contrast, a genetic optimiser behaves as Charles Darwin decreed and it throws off random mutations (aka sports) to see if they lead anywhere useful. So your chances of ending up on the summit of Everest and not K2 are greatly improved. The only practical downside is that this does take much longer. I have spent literally months of nightly runs on the PC in search of the very best answers. Having posed the right questions, at least you don't have to actually be there while it finds the answers! But exactly what are the right questions? If you don't ask the right questions, the answers will mislead you. It would appear there is some significant ambiguity out there.

TUNING FOR PERFORMANCE

The crunch question is, what exactly do you want to tune for? Very rarely do best Gain, F/R and SWR coincide. What I want is my Yagi optimised for good performance in every direction other than that of the station I am having a QSO with. It might sound strange put that way, but I know that there is almost nothing I can do which will influence the strength of the signal I can receive or transmit in the main lobe off the front. But off the rear, that is another matter. On 15m for example, I could indeed trade 0.05dB gain improvement off the front for 11dB degradation off the rear.

Contemplate that horrendous and typical trade-off. It is inherent in any and every 3- element Yagi.

I live near the NW edge of a small rural village. When beaming away from the village to the USA, my receiver noise floor is some 9dB better versus beaming into the village. This is far more important to me than any fractional dB gain improvement in the main lobe.

Further, much of the literature refers to the wonders of beam performance on transmit but mentions nothing about the mayhem you can cause to others if you tune for maximum gain at the expense of rear performance. And for what? Next to nothing. I regard it as precisely analogous to tuning your linear for 0.05dB more output – at the expense of 11dB increase in IMD products. Am I missing something here? I claim no expertise and I really would love to know.

There are indeed occasions where the facility to listen off the rear – and even at times to transmit off the rear would be useful. But that is a story for later.

In fact, PICaYAGI can store up to three tuning solutions per frequency so it is entirely possible to retain Best F/R and Best Gain as alternatives. The only time I use the latter is if the band is closed behind me and I'm working a station that is having trouble copying me. I can't say it gives much obvious improvement and, generally, I would prefer to suppress my local noise floor, there 24 hours a day. And I have a quiet location.

Finally in this context, there is one significant advantage we have over most of the antenna manufacturers. They cannot predict the customer's installed height or their ground characteristics. Since most of these offerings are designed to be used essentially out of the box they need to adopt a relatively low Q approach so that the antenna will deliver acceptable performance over a reasonable range. This is a compromise we don't have to make since, ultimately, this antenna is tuned from *your* shack at *your* installed height and over *your* ground. So we can indeed design for better monofrequency performance. Which is not only better, it is easier!

PICaYAGI MODELLING PROCESS

For a multi-band design the process is somewhat iterative. The first task is to find the best values for the relatively uncritical variables. That is, boom length and driven element spacing – which are indeed uncritical by comparison with the element lengths. This entails modelling all the variables over a plausible range of boom lengths and spacings for at least a few representative frequencies in each band, certainly including the bottom end of 20m and somewhere near the top of 10m.

If building a mono-band design, there are some well-established guidelines for boom length. So a good starting point is to choose a 3-element boom length that is about right for 17m, use it somewhat short on 20m and increasingly long on 15 / 12 / 10m. I finally settled for 4360mm (just over 14ft), mostly because I already had a suitable pole of that length.

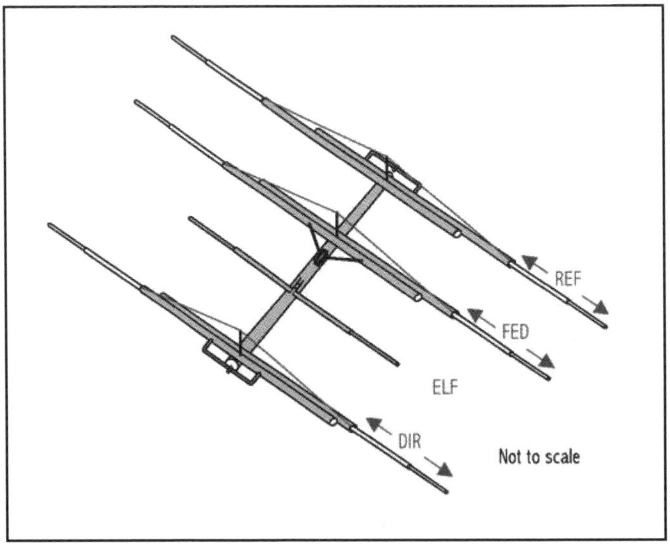

The other observation from the modelling is that the position of the driven element on the boom is not critical and anything from about 1/3 to 2/3 from either end is fine. The classic issue with a short boom is the narrow SWR bandwidth. But with the element lengths adjustable, this is a non-problem in our case. On the other hand, a boom that is too long gives poor F/R and there is no obvious way round this.

Much trawling of the web looking at other solutions gave

Fig 8.3: PICaYAGI general mechanical overview.

nothing helpful. For example, the FAQ on the SteppIR website **[5]** quotes 10 – 20dB F/R on 12m and 10m for the 3-ele design. Their brochure quotes F/R of 15dB on 12m and 11dB on 10m. This is with a 16ft boom – even longer than mine.

Because this is nothing like the sort of performance I wanted off the rear, I decided to add a fixed-length ELement Four (known as the ELF) – with switched stubs – for 12m and 10m only. This makes my 4360mm boom just about perfect for these two bands as well and, what's more, it delivers 4-ele performance.

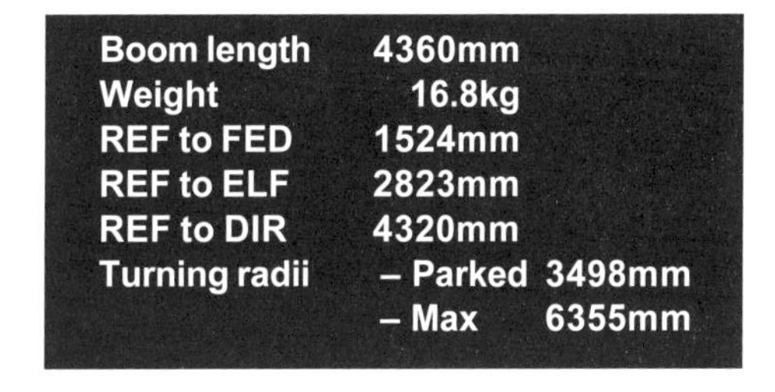

Boom length	4360mm	
Weight	16.8kg	
REF to FED	1524mm	
REF to ELF	2823mm	
REF to DIR	4320mm	
Turning radii	– Parked	3498mm
	– Max	6355mm

Table 8.1: Main mechanical parameters of the prototype PICaYAGI.

Having chosen the best boom length and element spacings, the final task is to determine if the elements would tune from the bottom of 20m to somewhere near the top of 10m. All the modelling indications were that it was distinctly marginal.

So a switchable linear resonator was added to both the main director and to the reflector to increase their effective maximum lengths by about 1.5%. That may not sound much, but it is 200kHz on 20m. I don't like linear resonators for large changes in resonant frequency because they radiate and potentially upset the pattern. They also have high circulating currents that can increase the losses. But a small one is harmless enough. By luck (not by design), the ability to switch the linear resonators on and off gives some valuable instant pattern switching options.

The final configuration is shown in **Fig 8.3**. This also defines the names of the elements, which I'm going to use from now on. (For the User Interface I need some distinctive and intuitive 3-letter abbreviations with a unique first letter). **Table 8.1** describes the main mechanical attributes.

You can also see the linear resonators on the REF and the DIR, the anti-droop cords on the three main elements and an early glimpse of the matching arrangements. More detail on that and the electronic aspects next month.

PART 2 – THE ELECTRONICS

This part covers the construction of the various electronic assemblies, principally the Command Unit that goes in the shack (and is constructionally trivial) – and the Controller, which mounts on the Yagi boom. The functional relationship between them is shown in **Fig 8.4**. This also shows the two user interfaces. During commissioning, a PC is used to control all the Yagi parameters via an RS232 link (max 100m). The outcome of this is up to three antenna configurations (or 'solutions') stored per frequency. During normal operation the Controller measures your instantaneous Tx frequency and following any change, the Command Unit is mostly used simply to tell PICaYAGI to go to one of those stored solutions. All the detail follows later.

COMMAND UNIT

See **Fig 8.5** and **Photo 5**. The purpose of the Command Unit is to route DC power up the coax to the Controller. The operation of SW1, the Command Switch, may look obtuse since it evidently does the same job in both switch positions. The secret lies in the brief break in supply during the changeover period. This break is detected by the PIC in the Controller and interpreted as a user command.

This simple circuit is built into a PCB box about 8 x 4 x 4cm. The best construction method uses point to point wiring that not only saves making a PCB but also keeps the strays to a minimum. It is very important that this unit be fitted last in the coax line up to PICaYAGI, since many SWR bridges, power meters and coax switches could be damaged if you apply DC power to them via the coax inner.

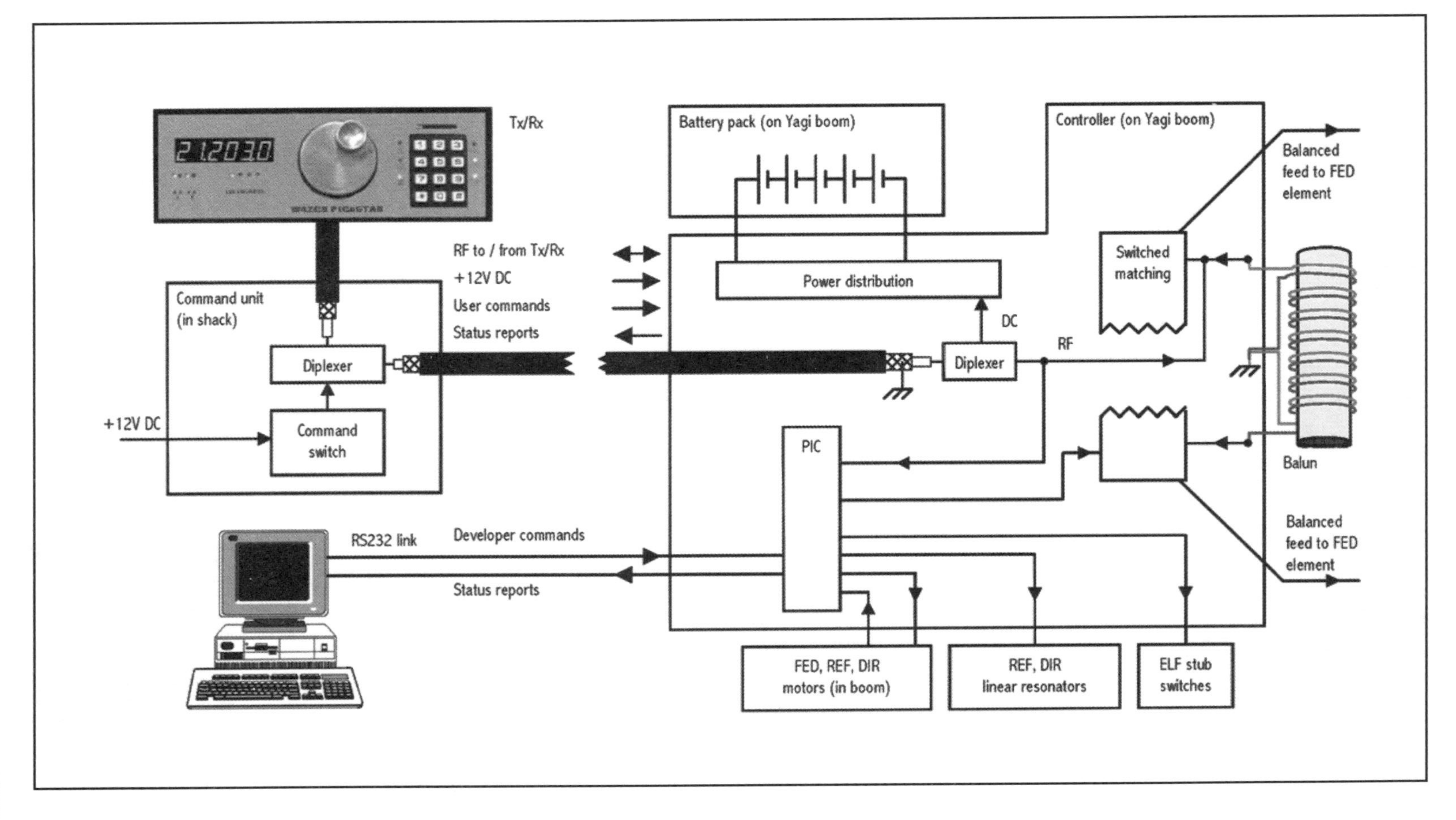

Fig 8. 4: Block diagram of the PICaYAGI Control system.

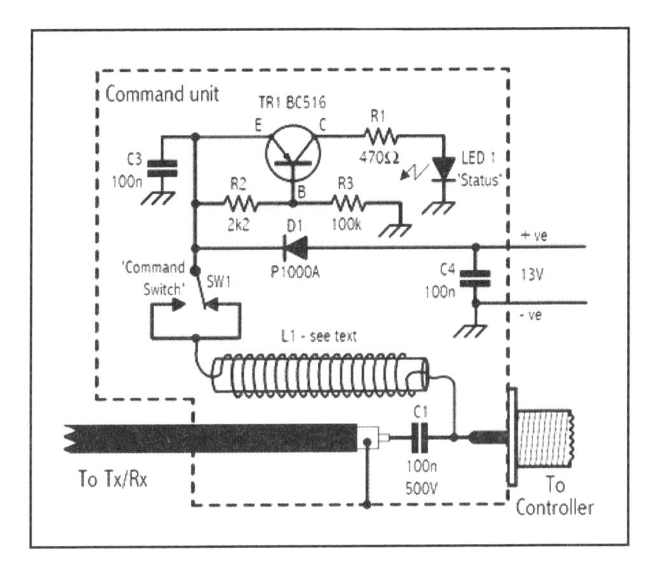

Fig 8.5: Command Unit circuit diagram.

Photo 5: Command Unit construction.

To prevent you from ever fitting the unit the wrong way round, I suggest you wire the coax to the Tx / Rx as a flying lead and fit a connector for the lead to the Controller, as illustrated.

CONTROLLER OVERVIEW

Fitted on the boom, this has the two logically separate functions of
1) controlling the PICaYAGI parameters and
2) providing matching between the single ended 50Ω coax and the 200Ω balanced feed to the FED element.

So it could have been built as two units but since neither would easily fit inside the boom it was felt better to keep them integrated for weatherproofing purposes.

CONTROLLER CONNECTORS

I haven't specified the connectors for the Controller or for the wiring out to the elements since there are many different strategies. An early decision is needed because it determines the Controller enclosure dimensions.

You could use *no* main connector on the Controller, solder wire tails to the PCB pads and then fit individual wire-wire connectors further out. Although more flexible, this approach does make the Controller more difficult to remove.

If you opt to fit a multi-way connector you need at least 28 poles. This allows two poles for each heavy current line – but only one pole for all the +5V feeds to the shaft encoders and one common pole for all their ground returns. The connector should be gold-flashed to give durability out there. I used a good quality 36-way shrouded IDC plug / socket (as used on parallel port printers).

For the individual wire connectors outside the boom, I used bullet connectors. These are available from all car accessory shops and used in the hostile environment under car bonnets. I crimped them with Mole grips. My thanks to David, G3YYD, for this suggestion.

Wiring to the motors and shaft encoders passes from outside to inside the boom via the smallest possible holes drilled in the underside of the boom. Since you need to be able to separate the boom modules for servicing, I confess to using the dreaded 5A terminal block ('choc block') inside the boom – because you need to remove the connectors from the wires to extract the wires through the boom wall.

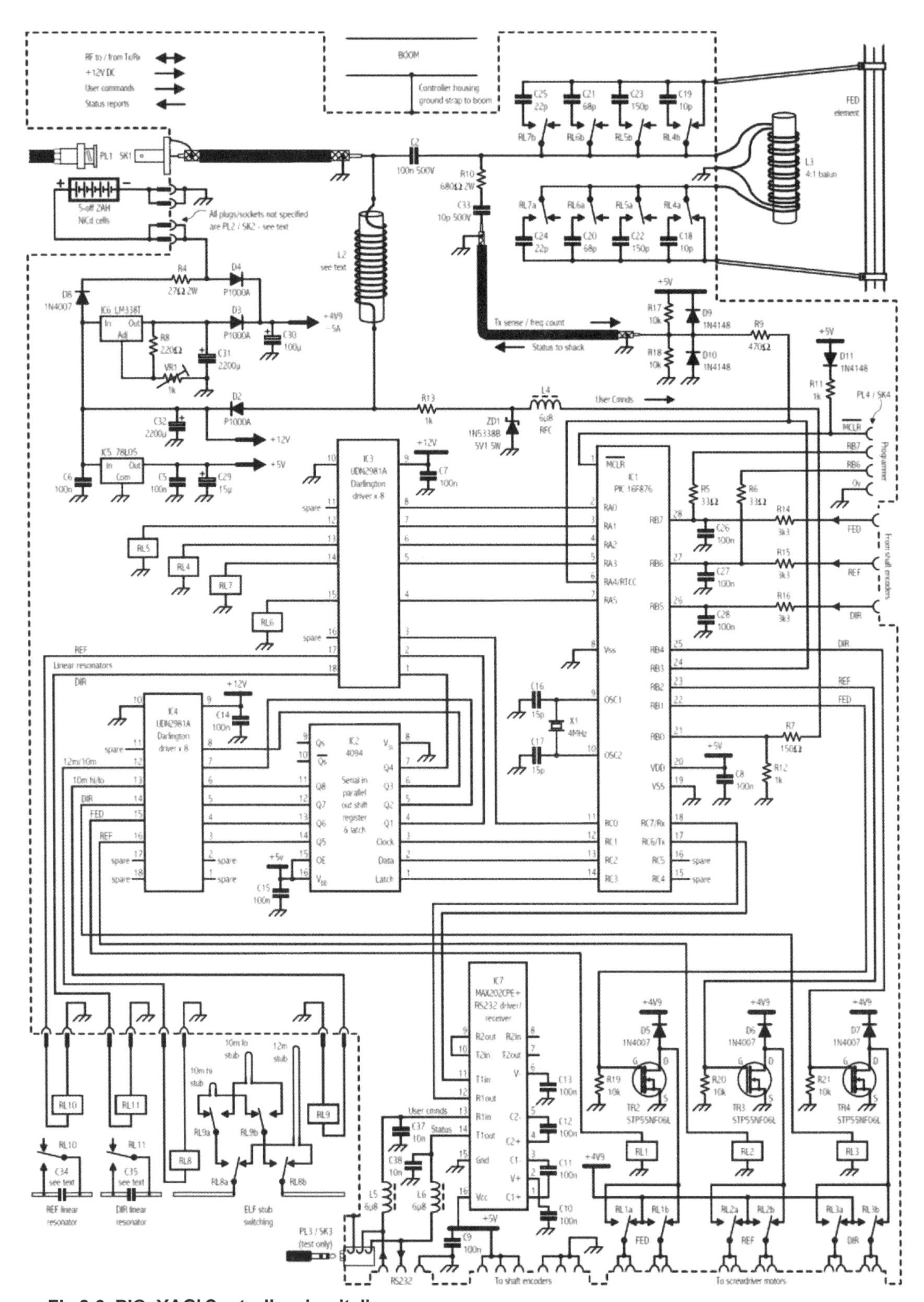

Fig 8.6: PICaYAGI Controller circuit diagram.

CONTROLLER DESCRIPTION

See **Fig 8.6**. RF and DC arriving up the coax are split by C2 and L2. The RF component is transformed to 200Ω balanced by L3 and then each arm passes via a series capacitor (determined by RL4 – RL7) and out to the taps on the FED element.

The DC component feeds the two regulators, IC5 and IC6 – and the presence of this DC is sensed by the PIC on the Usercmnds line. This Usercmnds line is DC coupled all the way back to your Command Unit and a brief break in the DC supply is used by the PIC to sense Command Switch operation. During these switch transients, power is maintained by the energy in C32, while D2 prevents reverse current flow.

Both the battery pack and IC6 supply power to the screwdriver motors for the elements. Most of the time, IC6 takes the load but if there is a lot of element movement in a short time, IC6 gets hotter, starts to shut down and the battery takes over. VR1 sets the point where this transition occurs. Charging is via D8 / R4. My thanks to Chris Stake for suggesting this approach and for the detailed implementation. It avoids the need for either a much larger capacity (and heavier) battery – or a much larger (and more expensive) regulator.

A sniff of RF is tapped off by R10 and C33, clamped by R17/18 and D9/10 and fed to the PIC to allow it to determine when the Tx is on. This line also goes to the PIC real time clock register via an internal 50MHz prescaler for frequency counting purposes. The values of R10 and C33 are somewhat critical in discriminating between no transmission and extreme QRP operation across the 20m – 10m spectrum while at the same time not saturating at higher power levels. CW status messages are fed back from the PIC to your Rx via this same line.

Relays RL1 / 2 / 3 define the direction of element movement – IN or OUT – for the FED, REF and DIR elements respectively. These are pre-set before applying power to the screwdriver motor via TR2 / 3 / 4. The hardware is configured so that, in principle, more than one element could be moved simultaneously. The extra complication in the software and the power budget issues are such that, at the time of writing, only one element is moved at a time.

The remainder of the PIC is devoted to relay switching, via IC2 / IC4 and IC3. There are enough spare lines that IC2 could be dispensed with, but I chose to retain the spare capacity.

My thanks to Bob, 5B4AGN, for suggesting the use of the hi-side drivers from the UDN family of ICs. These lead to simpler tracking than the more familiar (to me) lo-side drivers of the ULN family.

CONTROLLER CONSTRUCTION NOTES

This PCB is best manufactured by the iron-on toner transfer method. Many of us use this method nowadays and if you are not familiar with it, it is fully described in **[6]**. It gives no better results than photo-etch but requires less investment. If you normally wait for someone else to make PCBs for you I recommend you give it a go since the ability to make your own PCBs is the essential prerequisite of designing anything non-trivial yourself. This PCB is particularly easy because it is single-sided, with continuous unetched ground-plane on the non-track side.

I also commend to you a W4ZCB Beeper for checking out track shorts / opens and soldering integrity. Its great virtue is that, unlike some DMMs, it does not apply out of spec voltages to active devices. See **[6]** for details. You can build it dead-bug style in 30 minutes.

The Controller PCB is mounted – somewhat unconventionally – track side up. Components are mounted on both sides, as specified in **Fig 8.7**.

- All the large electrolytic capacitors are illustrated at approximate size and position.
- L2 leads are shown 'long' only for illustration purposes.

Most holes are drilled 0.7mm. For relays drill 1mm.

NB:-This image is mirrored - assuming toner transfer method of manufacture.

PCB dims 75 x 192mm (2.95" x 7.56" approx)

Top view of PCB (track side) showing component location.
Underside is unetched ground plane.

All ICs, relays, X1, SK1, C31, C32 and Tr2, 3, 4 are on underside.
All other components are track side.

Programmer interface SK4

MCLR

RB7

RB6

Ground

Lift one end of C26 and C27 to use this socket

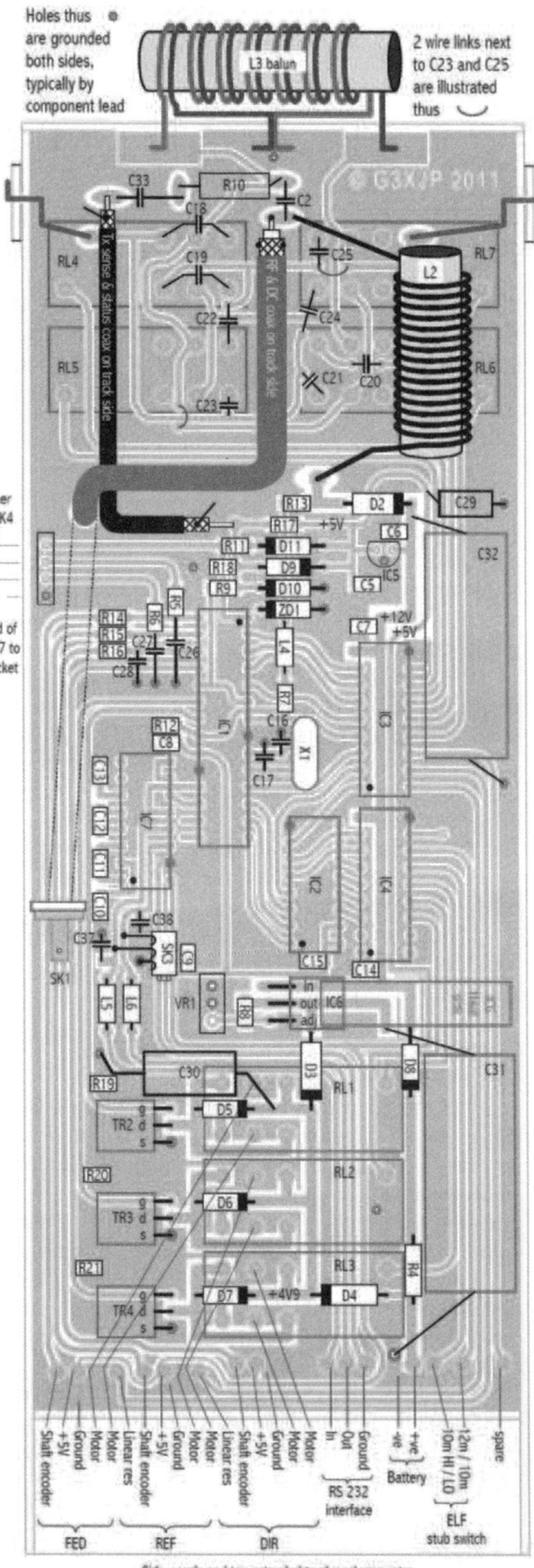

Fig 8.7: Controller PCB and component overlay. (NB: PCB track and drilling template shown at 90.1% of size: enlarge by 111% for full size.)

Photo 6: Controller from beneath before fitting the side and top plates. Note the end plate cut-out arcs.

Photo 7: Controller in PCB box with GRP shrouding mounted on the boom.

• Note that the balun must be fitted outside the box to avoid the box acting as a shorted turn on the coil.
• IC6 is mounted vertically. The extensive heat sink on IC6 is only partially illustrated. **Photo 6** shows the full detail of a finned heat sink and much aluminium bar bolted to IC6 and supported by PCB brackets.
• The main RF + DC coax is routed through the board to SK1 on the underside – mounted on a PCB bracket soldered to the side wall. This gives maximum shrouding to this critical connector. You need to cut a slot in the end plate to allow passage of the coax.
• TR2, 3, 4 are mounted vertically as shown in **Photo 6**. They don't need heatsinks.
• Programming socket SK4 is not essential; after the PIC is programmed for the first time off-board it can be re-programmed on the mast via the RS232 link.
• PL3 / SK3 are fitted to allow local testing of the Controller in the shack without using the main connector.

Photo 7 shows the completed assembly as fitted to the boom. It would be impractical to fully seal a box that has so many penetrations, so I chose the approach of sealing against rainfall from above but leaving it fully ventilated from below. To this end, the assembly is slightly wider than the boom diameter – and you can see a small ventilation gap adjacent to the boom on the lower long edge.

End plates. Determine the height of the end plates to give generous enclosure of all the PCB components. Then add a further 3cm. This allows you to cut out a partial circle (arc) so that the end plates sit astride your boom. If you are fitting a multi-way connector you may want to angle that end plate so that the leads from the connector go down towards the boom; in that case, that end plate will need to be correspondingly taller. And, at least in theory, the cut-out for the boom then needs to be parabolic (but not in practice!)

The Controller is mounted on the boom by fabricating GRP mouldings bonded on to each end plate, which are then secured to the boom by hose clips (that also retain the Controller's grounding strap). This allows the Controller to be moved along the boom to establish the best geometry for matching. The technique for making GRP mouldings is described shortly.

Side plates. The height of the side plates is the same as that of the end plates. Their length is determined by the need to shroud any connector(s). Each side plate has a small rectangular PCB 'washer' bonded to it to act as a feed-through insulator. 2mm wire takes RF from inside to outside the enclosure. Before fitting the side plate, drill the hole in the washer and side plate and countersink all but the outside face of the washer. Solder in the side plates, bond the washers to the side plates with epoxy resin and solder in the 2mm wire before the epoxy resin cures.

Top plate. This is not soldered to the others. It is extended with a thin layer of GRP to form lips on all four sides – and down and along to shroud any connector and the first few cm of wiring.

COMPONENT LIST

Capacitors (wire ended unless stated)

C1, C2	100n, 500V
C3, C4, C26-C28	100n disc ceramic
C5-C15	100n (1208 SMD)
C16, C17	15p ceramic
C18, C19	10p 350V silver mica
C20, C21	68p 350V silver mica
C22, C23	150p 350V silver mica
C24, C25	22p 350V silver mica
C37, C38	10n disc ceramic
C29	15µ 10V electrolytic
C30	100µ 10V electrolytic
C31, C32	2200µ 15V electrolytic
C33	10p 500V
C34, C35 (linear resonators)	see text
C36 (x3 for shaft encoders)	100n disc ceramic

Inductors

L1, L2, L3 are all wound on 5 x 50mm ferrite rod

L1, L2	chokes 19t of 2mm EnCu wire
L3, 4:1 balun	8 bifilar turns of 2mm EnCu wire
L4-L6	6µ8 axial choke

Resistors – wire ended

R1	4700
R2	2k2
R3	100k
R4	270 2W
R10	6800 2W
R22, R23 (x3 each for shaft encoders)	2k

Resistors - 1208 SMD

R5, R6	33Ω	R11-R13	1k
R7	150Ω	R14-R16	3k3
R8	220Ω	R17-R21	10k
R9	470Ω		

Semiconductors

D1-D4	10A, P100A
D5-D8	1N4007
D9-D11	1N4148
IC1	PIC 16F876
IC2	4094
IC3, IC4	UDN2981A
IC5	78L05
IC6	LM338T
IC7	MAX202CPE+
IC8 (x3 for shaft encoders)	KTIR0221DS
TR1	BC516
TR2-TR4	STP55NF06L N-channel power MOSFET

Relays

RL1-RL9	2p C/O, 240V mains rated, 12V coil
RL10, RL11	1p NO, 240V mains rated, 12V coil

Miscellaneous

SW1	1p momentary break, 5A
LED 1	not critical
VR1	1k, 10 turn preset
X1	4MHz wire-ended xtal
ZD1	1N5338B 5V1 5W Zener
5 x 2Ah NiCd cells with tags, soldered in series	
PL1 / SK1	50Ω BNC
PL2 / SK2 and PL4 / SK4	unspecified, see text
PL3 / SK3	stereo plug / socket for local testing
Motor M1	Homebase CSD3623 (3 off screwdrivers)

BATTERY PACK

This comprises five 2Ah cells soldered in line in series and fits neatly into some PVC waste pipe, which must be ventilated. It is taped under the boom; its position is used for fine tuning the boom's mechanical balance. I used NiCd cells rescued from an old cordless drill pack. I think NiMH would be fine since the charge rate is very low, but I haven't tried them.

GRP MOULDING PROCESS

For basic materials, I bought some general purpose resin, catalyst and fine glassfibre cloth via the web **[7]**, though you could use a car body repair kit for these. At the same time I bought two 1000mm lengths of 25.4 OD x 19.4mm ID GRP tube. This tube is used later for element fabrication purposes.

Several components are fabricated by GRP moulded to fit on to a particular shape, e.g. the boom, but without bonding to it. To achieve this, you must first cover that shape with something that the resin will not adhere to. Furniture polish or polyethylene sheet are popular resists but I find kitchen foil – shiny side out – gives the cleanest result. I used this approach for mounting the ELF relays and the

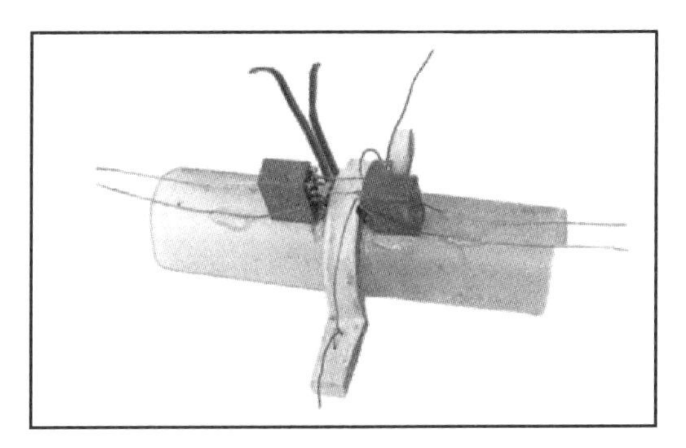

Photo 8: ELF stub relays RL8 and RL9 before encapsulation in GRP.

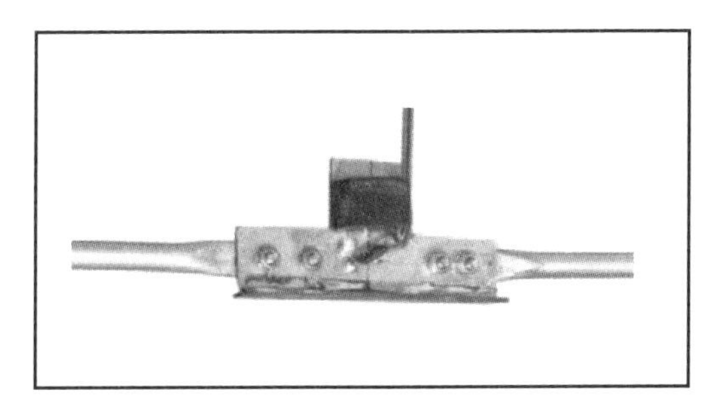

Photo 9: REF or DIR linear resonator relay mounting. RL10 is on the REF assembly, RL11 on the DIR assembly.

Controller to the boom and for extending the Controller top cover.

Form the kitchen foil to fit the shape (e.g. the boom), allowing plenty of excess at the edges. Fold, crinkle and pinch the edges to secure it or to add some rigidity. First paint on a layer of resin. Then add layers of fibreglass mat, saturating each layer with resin by stippling with a paint brush. Avoid brush strokes, which could move the glass strands. A large number of small pieces of mat is easier than the other way round.

Once you have enough mat thickness to retain its shape, let it cure, remove it and trim it with a pair of scissors. Then replace it and add further mat and resin to give the required strength, using the existing cut edges as a guide. There is always some resin left over, so I made a habit of painting yet another coat on the balun before cleaning up. Do not be tempted to apply resin to the PCB, as this makes any rework extremely difficult. I prefer several coats of clear acrylic auto varnish, which can be removed with a solvent – or even soldered through in emergencies.

ELF STUB RELAYS

As shown in **Photo 8**, I first laminated some GRP around just over half the circumference of the boom. I then bonded on the two relays, RL8 and RL9. You can see I also added some UPVC bar to support the ELF element. This was in the early days when I planned to use a wire ELF. You don't need that UPVC support bar.

Note that the 10m HI stub shown in **Fig 8.6** is a direct link between the RL9 terminals, i.e. the minimum possible length. Solder in that link. Bring out wire tails for the relay coils, the 10m LO and 12m stubs and the two half elements. Then completely encapsulated the assembly in GRP. This then clips on the underside of the boom, adjacent to the ELF element clamp.

LINEAR RESONATORS

One of these is fitted each to the REF and the DIR. See **Fig 8.8**. A normally open (NO) relay contact shorts the capacitor in the middle of the resonator to switch it off. The two resonators are constructionally identical.

As shown in **Photo 9**, the relay is bonded to a small piece of PCB, to which the tubing is bolted with 4mm stainless hardware. A further strip of PCB at right angles adds vertical strength. The PCB copper plane is simply scored up the middle to electrically separate the two halves.

A clamp is made from approx 330mm of 20 x 3mm flat bar by bending it round an element pair. Mole grips are used to force it closed around the element while drilling the screw holes and fitting the screws. You need 6 of these clamps in total – 4 for the linear resonators and 2 for the matching system (described later).

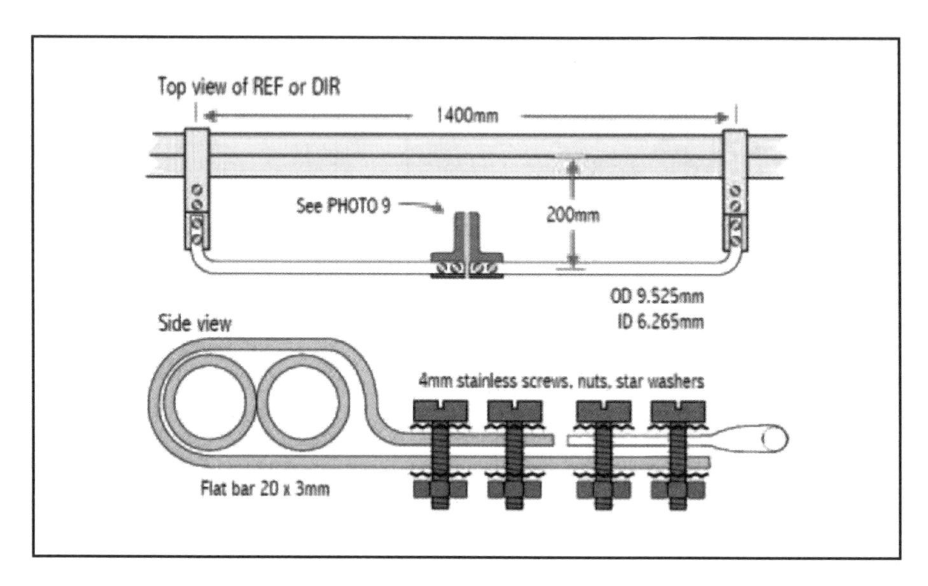

Fig 8.8: Linear resonator mechanical construction (not to scale).

The 9.525mm OD tube is bent to shape, flattened at both ends with a hammer, then filed even flatter and finally trimmed to length. For commissioning, use a trimmer of about 50pF for C34 and C35. I used mica compression types. These were subsequently replaced – after tuning – with 350V rating silver mica capacitors. Finally, the whole PCB assembly was encapsulated in GRP.

PART 3 – THE MECHANICS

This part describes the mechanical construction. **Fig 8.9** gives a perspective on the lengths and spacings and **Fig 8.10** shows a half element in more detail, emphasising the wall thicknesses, diameters and the dielectric coupling. The four tubes are

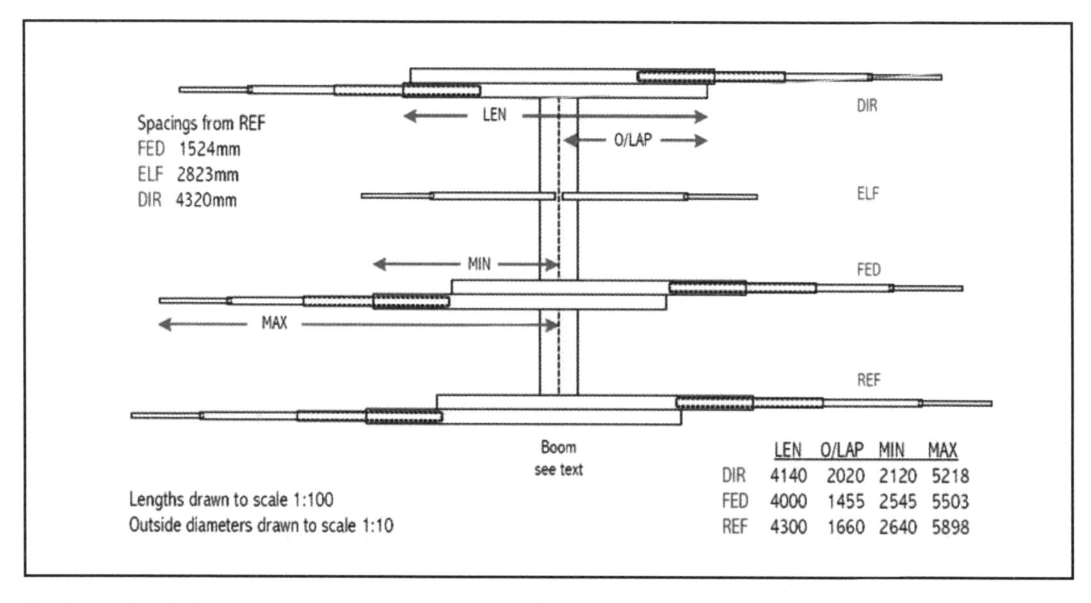

	LEN	O/LAP	MIN	MAX
DIR	4140	2020	2120	5218
FED	4000	1455	2545	5503
REF	4300	1660	2640	5898

Fig 8.9: PICaYAGI main element lengths – viewed from above.

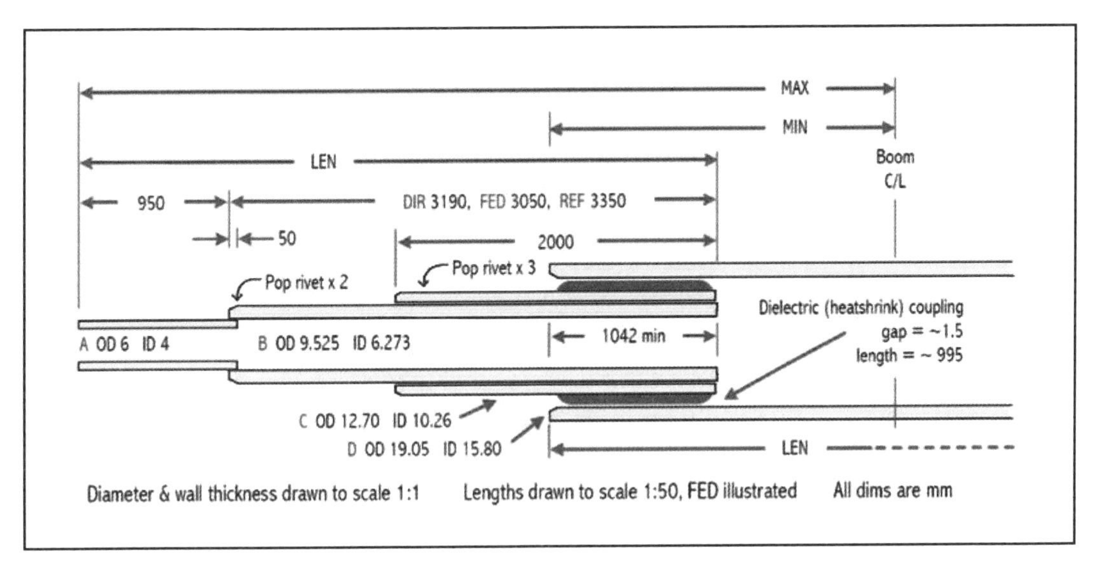

Fig 8.10: PICaYAGI half element at full extension, showing tube diameters and wall thickness at size.

lettered for reference. The table in **Fig 8.9** defines the differences between the three elements, i.e. the length (LEN) and overlap (O/LAP) of the fixed tube D. **Fig 8.10** defines the length of tube B by element. MIN and MAX follow by arithmetic.

MODULAR CONSTRUCTION

For the boom, I used 3 x 1500mm lengths of 1-7/8in OD tube with 16 SWG wall. Two of the lengths are swaged, which gives a net length of 4360mm – and plenty of rigidity with one exhaust clamp per joint. I bought this tube as a 'seconds' mast set. The 44.37mm ID is *just* enough to accommodate the screwdriver. But were I doing it again, I would buy 5000mm of 2in OD x 16 SWG and use some 2-1/4in OD x 16 SWG with exhaust clamps to externally sleeve the boom joints and the stub mast attachment point. See **Figure 8.11** for both strategies – and a summary of the drive train wiring that penetrates the boom.

The virtue of this modular approach is that each element can be built and tested in turn – and separated in the event of any problem. This is important in practice because although one such module might go in your garage (but for the other stuff), three joined together almost certainly won't. All my elements are secured to the top of the boom with standard TV boom / mast clamps.

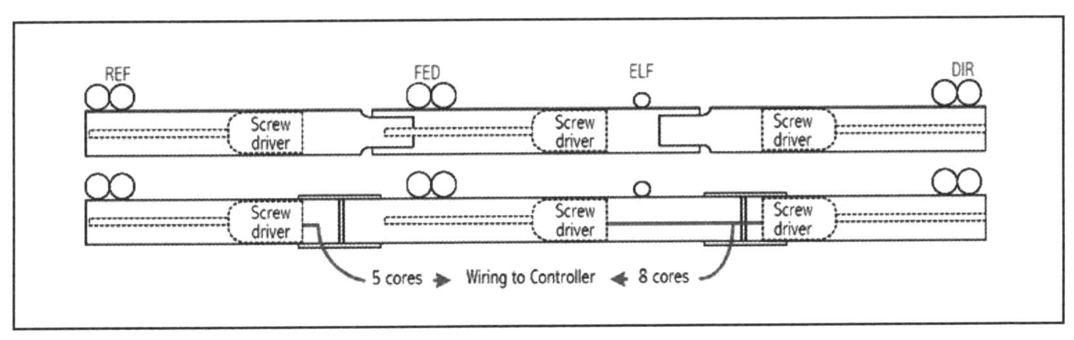

Fig 8.11: Boom and main element modular construction showing two alternative boom arrangements.

CONSTRUCTION SUMMARY

The basic sequence is as follows:

- Build 6 identical half-elements
- Assemble them into 3 pairs
- Attach them to their boom as REF, FED, DIR
- Build the 3 identical screwdriver assemblies
- Integrate them with their boom / element modules and add the braid drive cords
- Test each element for IN / OUT functionality
- Integrate the 3 boom / element modules
- Add the linear resonators and matching system
- Add the ELF
- Add the Controller – and you have your PICaYAGI.

Ah yes, not forgetting the software – this adds much to the functionality but little to the weight!

TUBE STOCK

This design makes efficient use of 4m or 5m standard length round tube stock. It is just capable of spanning 14.00 – 29.40MHz. At the time of writing, no one supplier in the UK stocks all the aluminium needed at a plausible price. There is a huge (e.g. 2:1 variation) in delivered aluminium tube prices available over the internet. However, the supplier I used two years ago **[8]** is still easily the best bet in the UK as of early 2011 – though unfortunately they don't stock quite everything you need. I note that they now sell some tube in 5m lengths, which gives more frequency span.

The structural element tubing, i.e. all but the tips, should be heat treated to 6063T6 or similar.

UNITS

I need to make an early apology for mixing dimensional units. In the 84 years since Yagi and Uda we have had 'standardisation' – as a result of which, most suppliers in the UK now sell aluminium tube in metric lengths with SWG wall thicknesses and Imperial diameters. Amazing! For purchasing consistency, I have elected to use these units – though they are certainly not the easiest to work with if you need to adapt for either pure metric or USA standard sizes.

ALUMINIUM SHOPPING LIST

Starting at the element tips and using the references in **Fig 8.10**:

A 9 off, 6mm OD x 4mm ID x 1000mm. From various web suppliers – but also sold in 1m lengths by most UK DIY shops. You need 8 x 1m element tips and ~1m for the matching unit.

B 6 off, 3/8in x 16 SWG x 5000mm (OD 9.525, ID 6.273mm). You need one per half element. The offcuts are used for screwdriver coupling, linear resonators, the ELF and the matching unit.

C 3 off, 1/2in x 18 SWG x 4000mm (OD 12.70, ID 10.26mm). 3 off – cut into 2 – gives 6 x 2000mm.

D 6 off, 3/4in x 16 SWG x 5000mm (OD 19.05, ID 15.80mm). You need one length (= LEN) per half element.

To complete the aluminium order you also need 1 off, 6mm dia rod x approx 1000mm, 1 off, flat bar 20 x 3 x approx 2000mm, 1 off, square tube 7 x 7 internal x 1000mm. You will also require a boom of your choice, see **Fig 8.11**. As of January 2011, shopping carefully and only from UK sources, the above came to about £160 including shipping and VAT.

BUILDING THE SLIDING TUBES

The initial task is to build 6 identical sliding inner tubes using as-supplied tube B length and no tips. It takes about 3 days – including glue and epoxy setting times. The process is the result of much trial and much error and re-work – and required several builds to optimise it. I'm convinced that the process is completely repeatable – mostly because I repeated it myself twelve times.

Both the detail and the exact build sequence are important. If you are using the same tube sizes then I commend it to you. If different, then I commend the principles and construction tips.

The result is shown in **Fig 8.12**.

Some comments about the ingredients:

Aluminium foil. The kitchen foil used here is from Lakeland™. It is very high quality, i.e. soft, malleable, thick and, critically, unembossed. It comes in 45cm and 50cm widths. I preferred 45cm. 50cm is useable but more difficult to handle.

Heatshrink. I used Rapid Electronics' 03-1125 that comes in 1.2m lengths. It has a K of 2.5, a 3:1 shrink ratio and 0.85mm nominal shrunk wall thickness. You need 6 off (plus some spares).

The remainder of the sliding tube is also shrouded in heatshrink to ensure it never touches the inboard tube at any extension. I used Rapid Electronics' 03-0346, which is specified as 12.7mm but just fits over the 12.7mm OD tube with a little silicone grease and lots of elbow grease. It also shrinks on to the 6mm tube tips. Fit this heatshrink before attaching the braid.

For shrinking, I used a hot-air paint stripping gun. This is somewhat excessive but is fine so long as you stay well back from the heatshrink and keep it on the move. It is also perfect for lighting real BBQs. If you need to buy something, I commend this as very much cheaper and more versatile than the official tool.

Silicone grease. You need to apply this liberally to all the rubbing surfaces, e.g. the outside of the heatshrink and the inside of the boom.

Nylon screws. These are cheese-head crews. You need both M3 and M4. I used stainless steel nuts where required. My thanks to Ray, G4TZR, for the idea for these

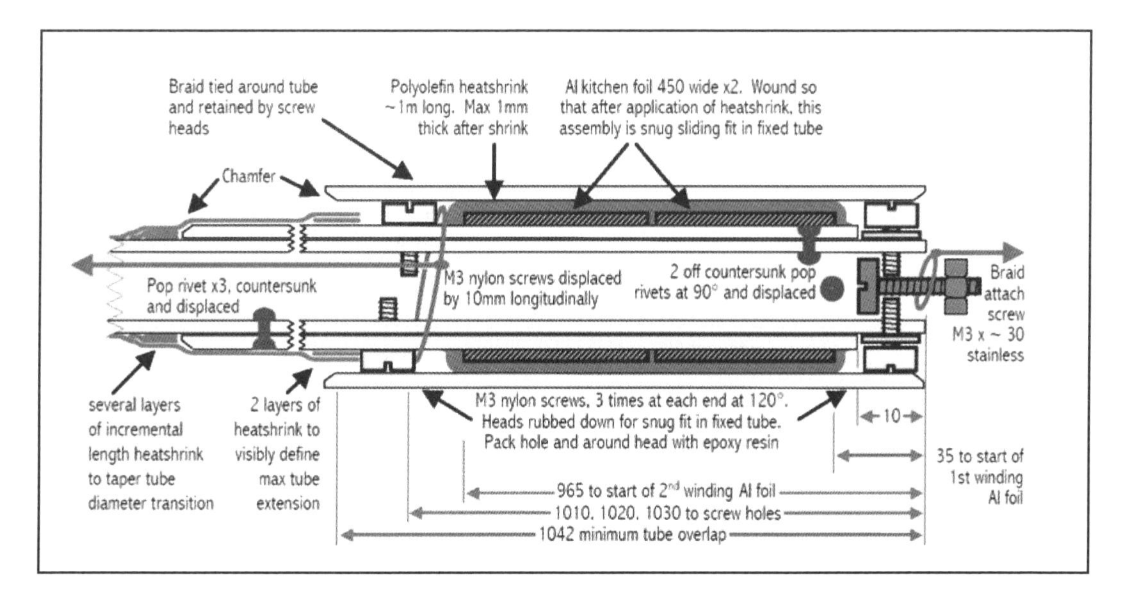

Fig 8.12: Details of assembly to form a dielectric-filled gap between fixed inboard and sliding tubes with minimal trapped air. Also shown are nylon bearings (M3 screws) to hold the sliding tube straight and true within the fixed tube. All dimensions are mm.

as bearings. I can see absolutely no evidence of wear after 2 years of hard use.

Twin pack epoxy resin. This is used to locate the nylon screws. It does not bond effectively to nylon but it does capture the thread and locate the head. Cleanliness is everything.

Pop rivets. These are countersunk type and 3mm diameter. You need about 40 of them in total. Keep the shanks for pinning the drive chain.

Penetrox™. This is an aluminium loaded grease by Burndy. It is used throughout this project on each and every metal-to-metal mating surface.

Braid. After much trial and error, I settled on Moss Green Fireline braid from Berkley. You need 80lb (36.3kg) breaking strain. A 300yd reel (way too much) costs about £28. I used cheaper 1mm Kevlar for the anti-droop braid, for which either will do fine.

INTEGRATING THE PAIRS

Measure the capacitance between the 6 fixed and sliding tubes to give 3 closest match pairs. The pair with the most capacitance should be assigned to the FED, followed by REF and then DIR in that order. The tubes can now be cut to length and the tips fitted. Next, some GRP fabrication. This is based on the glassfibre tube 25.4 OD x 19.4 ID x 1000mm mentioned in **Part 2**. This material is not pleasant to machine. Wear a mask and gloves and work outside if possible. It is used for three purposes:

- Tips for the inboard tubes
- Centre insulator for each main element
- Centre insulator for the ELF, discussed later.

GRP tips. There are 2 per element, bonded to the outer ends of the fixed inboard tubes (see **Fig 8.13**). These tips have three purposes:

1) to locate two M4 nylon screws that act as a support for the sliding tube,
2) to retain the anti-droop braid ends and
3) to retain some 5mm stainless steel bar as a captive roller, which is used to reverse the braid direction.

Cut 6 lengths of GRP tube 60mm long. For the OUT-braid tip, the roller goes on the bottom but for the INTER-braid tip it goes on the side. The alternative positions are illustrated in **Fig 8.13** – but this is only determined when the nylon screws and tips are actually fitted.

Hacksaw and file a tapered slot to capture the roller in the thickness of the GRP wall – just protruding into the ID of the GRP tube. Then file a further notch to pass the braid.

Main element centre insulators. First, you need to cut along the length of the GRP tube with a hacksaw, tangential to the ID as in **Fig 8.14**. When finished, you need 7

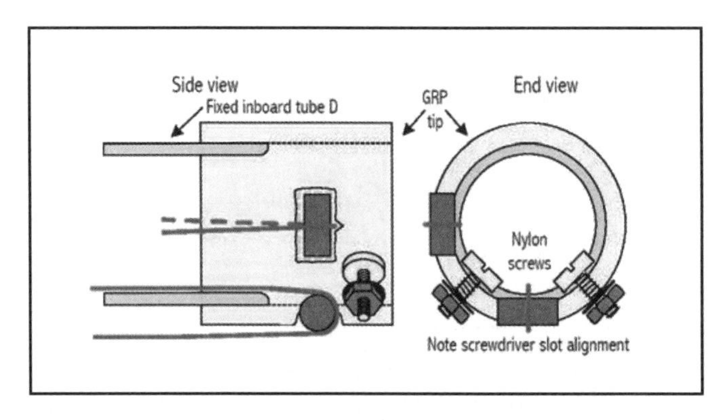

Fig 8.13: GRP tip showing both roller positions.

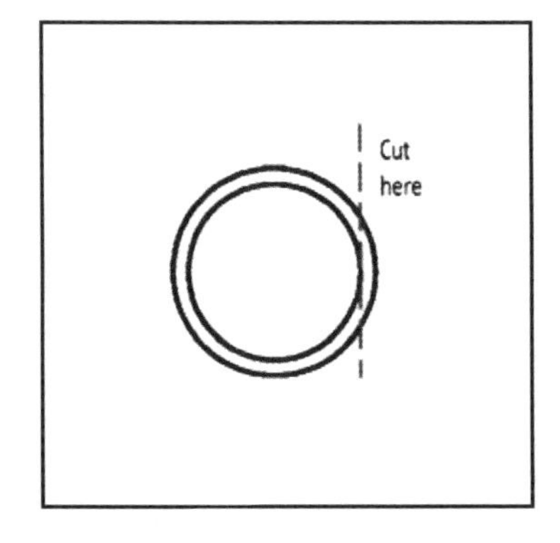

Fig 8.14: Cutting the GRP tube.

lengths each about 150mm long. The best way I could find of doing this by eye is to start the cut and then use this to draw two parallel lines on the outside to act as cutting guides. When you have the 7 lengths, set aside 1 for the ELF centre insulator and pair up the others. Each pair forms a central shroud over the middle of a main element to insulate it from the boom – and to prevent the boom to element clamp from crushing the element.

Photo 10 shows the finished result. Using two tubes as templates, apply some polish to the aluminium to act as an epoxy resist, slide on the 2 halves, clamp it all in a vice and stick the 2 halves together with a minimum of epoxy resin. When this has cured, check that it releases before applying layers of GRP mat and resin to integrate the two halves. At the same time, you need to integrate a vertical rod to hold up the centre of an anti-droop bracing cord. I used 15mm hardwood dowel about 350mm long with a V-notch to retain the braid. Wrap the dowel with a layer of mat and then resin and, when it has part-cured, bond it in the middle of the tubes. Then strengthen the joint with more GRP, tapering it off over the first few cm of the dowel. This process is best completed with the boom to element clamp in place, laminating around it.

Photo 10: Showing a central insulator with its vertical support for the anti-droop braid.

Grease the inside of the GRP and insert the tubes, adjusting the overlap. Clamp them together with stainless steel hose clips, 6 per element, with 1 hose clip each side of the GRP to prevent the tubes from pulling through. Then fit the GRP tips and their nylon screws, rubbing down their heads to wedge-shaped along the direction of element movement – and just proud of the ID of tube D.

MOVING THE ELEMENTS

The FED, DIR and REF mechanisms are identical. **Fig 8.15** shows the scheme of things. There are 3 braids that all have their directions reversed near the ends of the

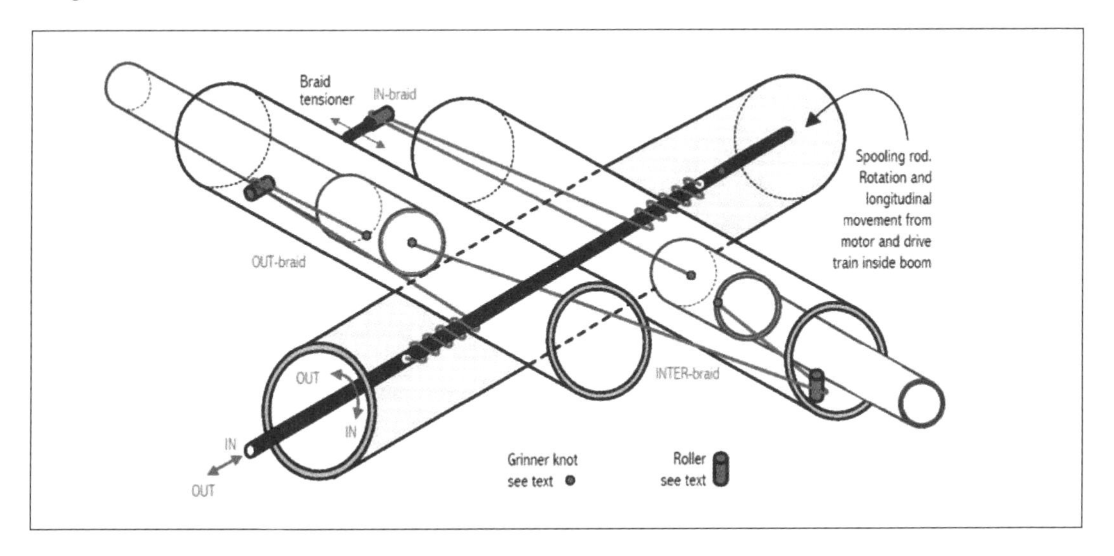

Fig 8.15: Overview of braid arrangements for pulling both sliding tubes IN or OUT.

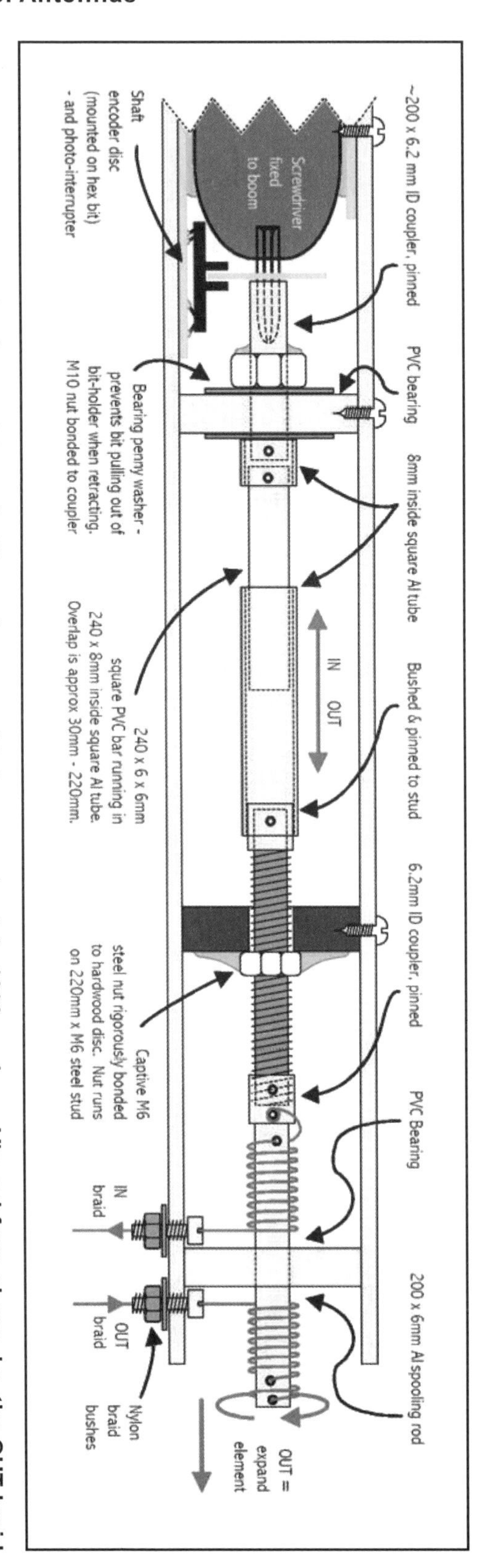

Fig 8.16: Spooling mechanism, not to scale. The entire assembly is approximately 1000mm long. Viewed from above, i.e. the OUT braid passes under the spooling rod.

fixed tubes. To expand the element, the OUT-braid is spooled on to the spooling rod and the left hand element goes out. The INTER-braid transfers this out motion to the right hand element, which also moves out – and takes up the IN-braid slack
as it is spooled off. And vice versa to retract.

The spooling rod is 6mm diameter, 18.85mm circumference. So to move the element by 3000mm requires about 160 turns. To avoid the braid piling up and altering the effective diameter, the spooling rod is driven longitudinally by an M6 thread. This has a 1mm pitch and so travels 160mm for 160 turns. The drive mechanism is shown in **Fig 8.16**. The essential feature is a sliding square drive. My thanks to Dave, G3SUL, for the idea. It is implemented as a square PVC rod sliding in a square aluminium tube. That PVC rod and the two PVC circular bearings are cut from a self-healing kitchen chopping board – available from all good supermarkets. None of the lengths is critical; there's no problem if everything is a bit longer than theoretically required.

SCREWDRIVER ASSEMBLY

This comprises a screwdriver motor and hex bit, a photointerrupter IC, a few components and some PCB tracking. **Fig 8.17** shows the electrical and mechanical detail. Detailed construction is not critical.

Fig 8.18 is a template for the encoder disc, which is made from PCB. Cut out the 28mm dia circle somewhat oversize. Drill the hex hole at 6mm and then file to hexagonal. Fit to a hex bit, mount it in a drill and turn the diameter down to size with some coarse emery cloth. Cut the 32 radial slots with a junior hacksaw. Remove all swarf.

Fit a Phillips screwdriver bit to a ~200 x 6.27mm ID aluminium tube by banging it in with a hammer. Ensure there is just enough of the bit showing to carry the encoder disc and fully insert it into the screwdriver. Bond the disc to the hex bit with epoxy resin. Two PCB strips about 10mm wide are bonded to the case of the screwdriver on opposite sides to locate it and to give an easy sliding fit in the boom. One of the PCB strips is scored (or etched) to provide three longitudinal tracks on which to mount the components and to interface the wiring back to the Controller.

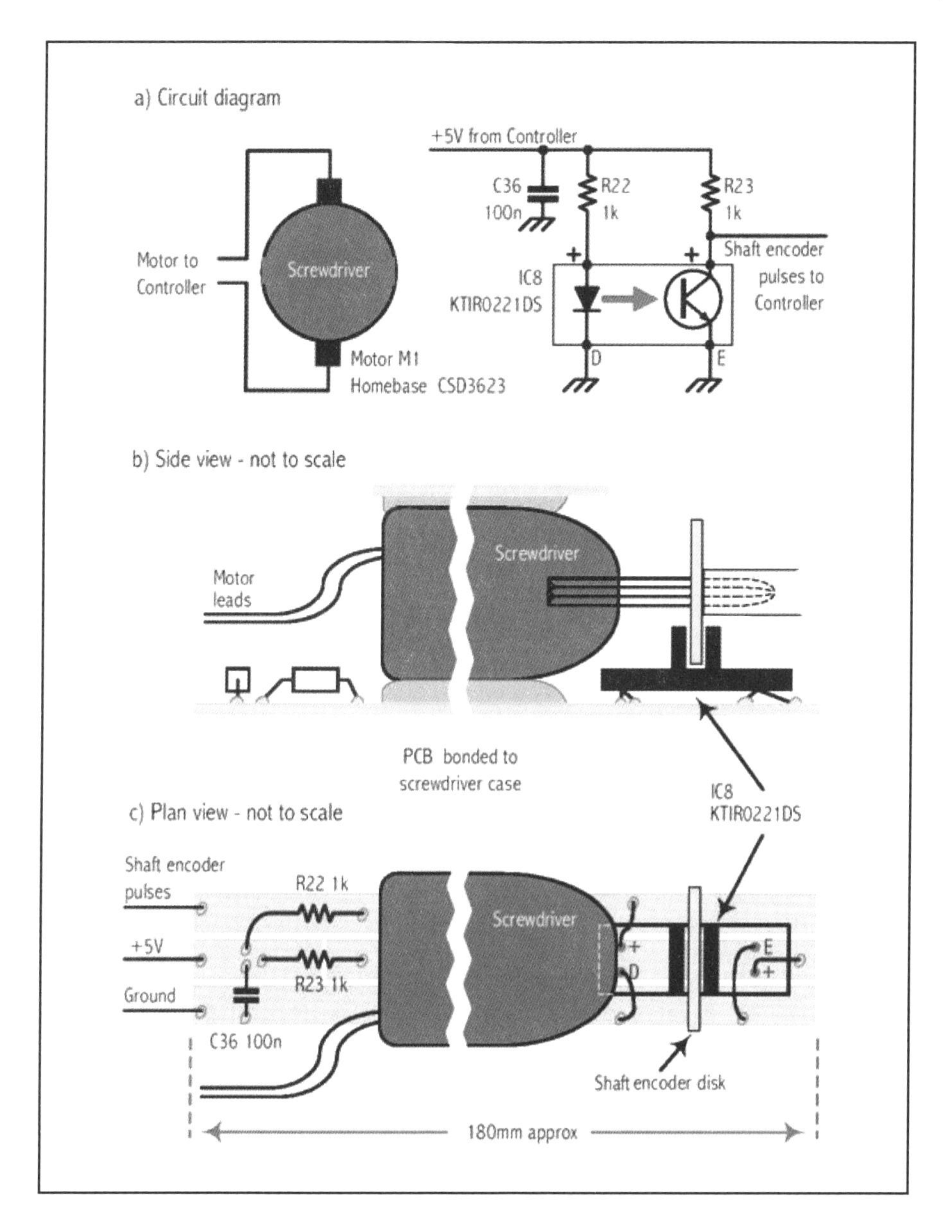

Fig 8.17: Screwdriver electrical and mechanical details.

Fabricate the PCB tracking. Then remove the battery pack, switches etc from the screwdriver and solder two generous 5A wire tails to the motor wiring. The recommended screwdriver is far from a round section, so trim off any surplus style bumps etc from the casing until a) it is a sliding fit inside the boom with the PCB strips and b) the motor axis is central in the boom.

Bond the screwdriver to the PCB with epoxy resin while sliding it into the boom with the bit and disc to check that the axis of the screwdriver is centred in – and parallel to – the boom.

Solder in all the components, adjusting the leads of the photo-interrupter so that the encoder disc sits centrally in its slot.

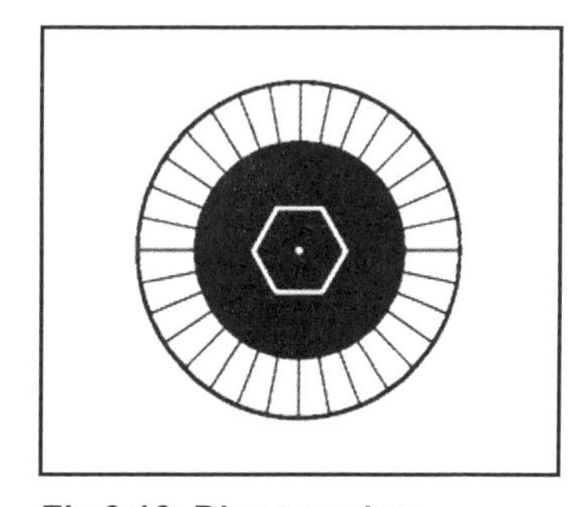

Fig 8.18: Disc template.

INTEGRATING DRIVE TRAIN, BRAID AND ELEMENT

Starting with an element pair lightly clamped to its boom section and a completed screwdriver drive train, the task is to couple these up to the IN, OUT and INTER-braids, ending up with a fully working expandable element, as shown in **Fig 8.15**.

The first task is to learn to tie a 7-turn Grinner knot. It is a popular fishing knot that minimises the reduction in braid breaking strain inherent in tying any knot. There are countless descriptions on the web, though I suggest you avoid the videos and find a step by step illustrated sequence. Some say you should use cyanoacrylate glue to secure the knot. For this application you don't need it and, because it is a slip knot, you can loosen and remove it without undoing it.

Another early task is to pre-stretch some braid. Hang up a long length and load it to about half breaking strain. A bucket full of water and some house bricks is useful. Leave it at least overnight. But be very careful where you put it because this braid is essentially invisible.

Nylon braid bushes. Drill a 0.7mm hole up the middle of two 10mm long M4 nylon screws. This is very much easier than it sounds. With their nuts and washers, these become the nylon IN and OUT braid bushes shown in **Fig 8.16**.

The boom needs two holes to locate the bushes, which go under the fixed element tubes. They are 19.05mm apart horizontally (a fixed tube D diameter) and mutually displaced vertically by about 4mm. Mark this off on the boom and drill. The element – and element to boom clamp – are subsequently rotated to 'point' these bushes at the top and bottom of the spooling rod.

Braid tensioner. This is used to take up any slack in the IN-braid. It is made very simply from a standard PVC wall mounting pipe clip for 22mm central heating pipe. This is one of the rare but happy occasions where they package them in threes and that is what you want. It clips to the element tube – and a length of 5mm roller (same as the GRP element tips) fits into what is normally the hole for screwing the pipe clip to the wall. The body of the clip may need to have some protruding flanges shaved off so that a hose clip fits neatly around it and sits flat. See **Photo 11**.

Drill a 1m hole both sides of the roller hole and also remove the PVC behind the roller hole so that the braid will go in, round the roller and back out again. Fit the braid first, then the roller. Check that the braid runs freely and is not trapped. Then clip it to the boom and tighten the hose clip to secure it. At any time its position may be adjusted to take up any slack in the IN-braid and also modest slack in the OUT-braid.

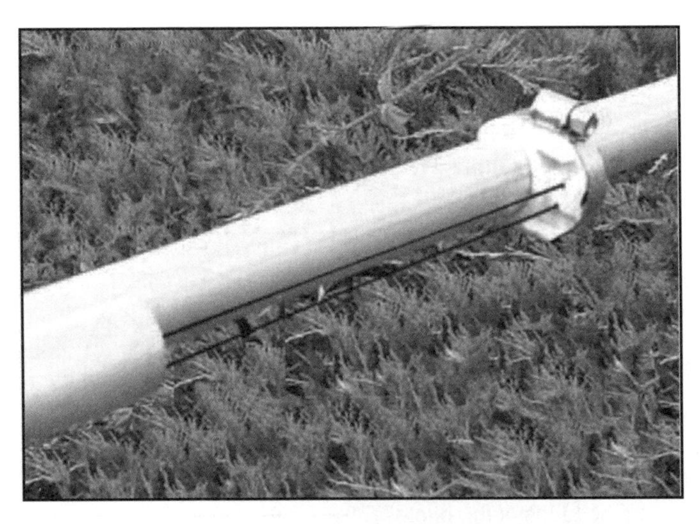

Photo 11: Braid tensioner made from a PVC pipe clip.

Fitting the anti-droop braid. Fully retract the elements. Using a Grinner knot, tie the braid around the element GRP tip using the bearing screws to prevent it sliding back along the tube (see **Photo 12**). Repeat at the other end, leaving the braid just slack. When you now raise the middle of the braid and slot it into the V-notch it should just bow the fixed tubes slightly upwards. Putting it another way, getting the tension right is a bit trial and error. But you can easily shorten the braid by sliding one end off the tube and then looping it through a steel nut (a few times and / or through several nuts) and re-attaching it.

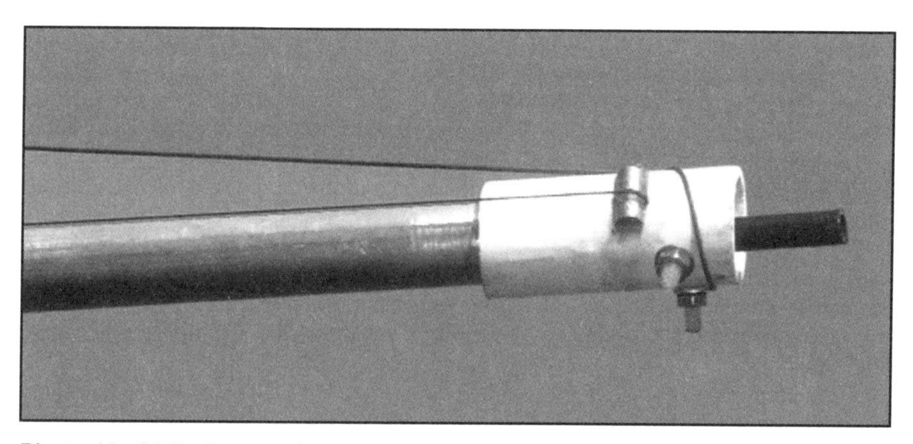

Photo 12: GRP element tip showing the two nylon screw bearings, anti-droop braid above – and INTERbraid with roller on the side.

Locating the drive train. Drill the hole in the spooling rod for the IN-braid. Drill it 1mm and then, hand-holding a 2mm drill bit, soften the sharp edges on both sides. Repeat with a 3mm drill bit. Do not drill the OUT-braid attach holes at this stage.

Rotate the M6 nut on its hardwood disc until it is 10mm from the screwdriver end of the stud. Lay the drive assembly next to the boom and position it so that a) the INbraid spooling hole is opposite the IN-braid bush hole in the boom and b) the sliding drive has 170mm of exposed PVC drive rod showing. The drive train is now in the fully OUT position, with some margin for error.

On the boom, lay off and mark the positions of the securing screws for the hardwood disc, the PVC bearing nearest the screwdriver and the screwdriver itself. Select 6 short, sharp, tapered, self-tapping fully threaded wood screws. Drill holes in the boom at a diameter that will take the full diameter of the screw; 1 for the screwdriver, 2 for the PVC bearing on opposite sides of the boom and 3 for the hardwood disc, approximately equally spaced around the boom.

Secure some coarse emery to a broom stick and use it to remove any internal rough edges or swarf from all the holes inside the boom. Lightly grease the end of the inside diameter of the boom and insert first the wiring, then the screwdriver into the boom. Now grease also the circumference of the PVC bearing, the hardwood disc, the M6 stud thread and the spooling rod.

Insert the assembly by pushing on the spooling rod. A word of caution: never pull on the wiring or from the screwdriver end since you risk pulling the bit out of the bitholder and smashing the shaft encoder disc.

As soon as the hardwood disc has passed the IN-braid bush hole, fit the bush, washer and nut and loosely tighten. Pull the spooling rod out a little, pass the end of the reel of braid through the bush, through the spooling rod IN-braid hole and fasten it to the coupling pin with a Grinner knot.

Continue pushing the assembly into the boom until the PVC bearing reaches its screw holes. (Once the screwdriver has cleared, put the shank of a pop rivet in the hole so you can see when the PVC first arrives.)

Fit the screw in the screwdriver casing but only 1 of 2 in the PVC bearing. Now pull the spooling rod back out until the hardwood disc lines up with its screw holes. Fit only 2 of those 3 screws at this stage.

Drill and self-tap the outer PVC bearing to take a couple of long large wood screws, but don't fit them. They can be fitted later to give something to pull on should you ever need to extract that bearing. Fit that outer PVC bearing, pushing on it until it reaches the IN-braid bush. This bearing also has the secondary use of helping to prevent crushing of the boom by the element clamp. Now fit the OUT-braid bush – again, loosely tightened.

Fitting the drive braids. Lay off the IN-braid length – via the braid tensioner – to the attachment screw position on the inner end of the fully extended right hand tube. Cut the braid length.

Push the end of the braid up through the hole of the OUT-braid roller on the left hand tube and then out of the tube end. Pull out the sliding tube until it is just more than fully extended and with the bushes showing. Using a Grinner knot, tie on the OUT-braid as per the left hand end of **Fig 8.12**. Then push the element all the way back in and a bit more – until it is sticking out of the far end by about 10mm. Lay off the OUT-braid to the spooling rod and with a generous 200mm further allowance, cut that length.

Push the end of the braid through the hole of the INTER-braid roller on the right hand tube and then out of the tube end. Pull out the sliding tube until it is just more than fully extended and with the bushes showing. Using a Grinner knot, tie on the INTER-braid as per the left hand end of **Fig 8.12**. Then push the element all the way back in and a bit more – until it is sticking out of the far end by about 10mm. Lay off the INTER-braid to the attachment screw of the other inner tube and with a generous allowance, cut the braid length.

All 3 braids are now attached at one end only. With both tubes fully retracted, attach the IN-braid via the braid tensioner. With both tubes protruding by some small but equal amount, attach the other end of the INTER-braid using a Grinner knot – as per the right hand end of **Fig 8.12**. I used a stainless steel nut and no washer. Get the braid taut using the nut to alter the tension, taking any gross surplus as turns through a steel nut as necessary.

Now pull out both elements to full extension and take up *most* of the IN-braid slack using the braid tensioner. Leave 60mm of slack length.

You now need the screwdriver motor, for which you will need some electricity. I used a 6V battery with a 10A series diode. A 5V 5A PSU would be fine, or even 4 of the batteries that came with the screwdrivers (if you can arrange to charge them). Rotate the element around the boom until the IN and OUT braid bushes are 'pointing' at the top and bottom of the spooling rod respectively. Then tighten the element to boom clamp.

Reel in the IN-braid and both elements using the motor. You *must* verify that the spooling rod is rotating clockwise, so just dab the power on briefly first time. Reel it fully in – and about 10mm more. Take this opportunity to re-adjust the length of the INTER-braid as necessary.

Photo 13: OUT-braid – spooled via braid bush.

You now need to drill the hole in the spooling rod to attach the OUT-braid. See **Photo 13** for the finished appearance of the OUT-braid and its attachment to the spooling rod.

The hole goes opposite the nylon OUTbraid bush but 5mm further out. Mark the spooling rod where the hole is to be drilled. To gain access to drill it, drive the motor out while manually pulling the surplus IN-braid from the boom. When you have enough room to work, drill the hole 1mm diameter – and a further hole about 20mm nearer the end of the spooling rod. Remove sharp edges both sides of both holes and cut off any surplus spooling rod length.

Pull out both elements to take up all the IN- and INTER-braid slack. Drive the elements in again until they are once again over-retracted by 10mm.

Thread the free end of the OUT-braid through its bush, then wind on 4 turns anticlockwise as tight as you can – then through the first hole in the spooling rod. A pair of long-nosed pliers or tweezers is useful. Pull the braid taut and tape it to the spooling rod. Drive the elements out once again until you have enough access to secure the OUT-braid properly.Wind any surplus braid tightly between the holes and pass the braid through the second hole twice. Whittle a match stick to secure the braid in the second hole. This match is only replaced with multiple overhand knots when all the lengths are fully stable.

Exercising the element. First fit the remaining self-tapping screws in the boom and tighten them all. You now need to drive the element fully IN and fully OUT about 15 times until the braid lengths stabilise. This is mostly just knots tightening and bedding in.

Make sure you never drive the elements further out than fully out!

After every OUT, take up any slack in the braid tensioner. After every IN, check the length of the INTER-braid. This carries half the load of the other two braids so tends to settle down sooner.

After about 5 cycles and with the elements over-retracted by 10mm, slacken off the braid tensioner, pull both elements out to give some slack in the OUT-braid, remove the match stick and pull through about 10mm more OUT-braid and resecure it. Readjust the braid tensioner.

THE ELF

The middle of this element is made from two 1700mm lengths of 9.525 x 6.273mm tube and a centre insulator. The centre insulator is approximately 150mm of the sliced GRP tube.

Flatten one end of the aluminium tube. Drill and attach some multi-strand insulated wire tails with stainless hardware. Repeat for the other half. Place them in the sliced GRP tube with a small gap between, seal the ends temporarily and stuff the GRP tube with mat and resin.

The ELF tips are each a 1m length of 6mm OD, 4mm ID tube, which telescopes for adjustment. Slit the inners and secure with size 00 hose clips.

Fit the element to the boom with a standard TV clamp. Clip on the ELF stub relays adjacent. Solder the element tails to those from the stub relays.

DELTEE MATCH

To my knowledge, this matching system is something new and very useful. It looks like a cross between a Delta match and a Tee match (see **Fig 8.19**) – hence the name, DelTee match (or δT-match for short). It is the only way I know to feed a Yagi element that cannot be split over 5 bands (i.e. more than an octave). I arrived at

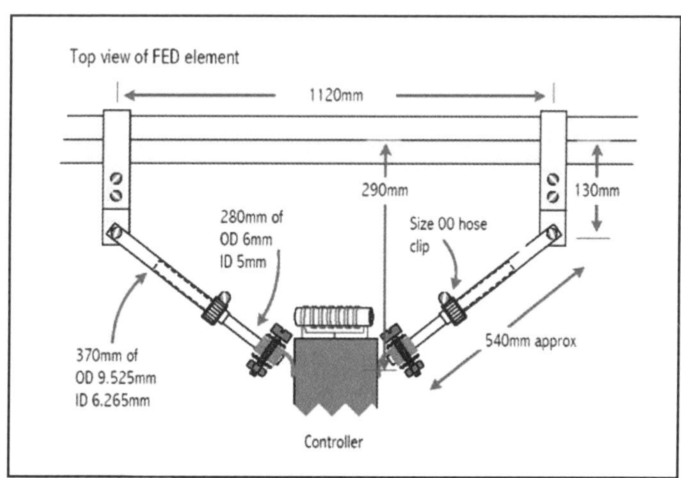

Fig 8.19: DelTee (δT) match construction. Not to scale.

Photo 14: Mechanical integration trials. Elements are fully parked.

the design by much tentative NEC modelling and then even more trial and error in practice.

The δT-match arms are screwed to some phosphor bronze strip (draught excluder or fingering) that in turn is soldered to the Controller feedthrough washers.

I'm not entirely sure exactly how it works so if you have any ideas, I would love to hear them. My own thoughts are as follows. Having tuned the Yagi on any given frequency for every aspect of performance except SWR, there remains a reactive and a resistive component to remove. Both these components can be removed by altering the length of the FED and changing the value of the series capacitors (in the Controller). This requires all the other element lengths to change as well to maintain the performance, so the process is highly iterative. There are several combinations that give a good match and some sort of performance but, according to NEC, only one combination that delivers it all.

Over the whole frequency span, the result is a Yagi that – if its driven element were split – would have a natural feed impedance between about 20Ω and 60Ω. Not split, it presents 200Ω at the balun terminals. There is very little variation in Yagi performance over that impedance range.

Photo 14 shows the first time the complete PICaYAGI was assembled for trials. The car gives a feel for scale.

PART 4 – SOFTWARE, USER INTERFACES, PERFORMANCE

This final part outlines the User Interfaces, how to tune your PICaYAGI and concludes with the measured performance results.

USER INTERFACES (UI) SUMMARY

As mentioned in previous parts, there are two UIs: 1) a Developer UI, to pre-define and save PICaYAGI tuning Solutions by frequency and 2) an Operational UI, for everyday use, which is mostly used simply to recall and apply those Solutions. One significant attraction of the Operational UI is that you don't need a PC for day-to-day operation.

At any one time you can choose to have either or both UIs in use. Further, you can start up the Developer UI part way through an operating session and it will seamlessly pick up the current state of play. This is particularly useful if you suspect something may have gone wrong.

The current state of all the PICaYAGI variables is stored in the Controller. So are all your tuning Solutions, indexed by frequency.

GENERAL DESIGN FEATURES

Before discussing the operation of the UIs, I need to explain some general design features for context.

Park. This an important occasion, when the element is fully retracted – because it is also when the element length gets recalibrated to zero.

As the element comes IN towards a mechanical end-stop it slows right down in anticipation until it finally gently bumps into the stop. It then goes back OUT again until the time per shaft encoder slot increases, which corresponds to any backlash in the OUT-braid being taken up. This point is defined as zero extension – and is highly repeatable. My thanks to Peter, OE6ZH, for inspiring this approach.

You need to fit a mechanical end-stop to each main element. Its purpose is both to prevent over-retraction and to define zero element length. The end-stop is simply a 4mm screw inserted in a hole drilled in the element very close to the end of the tube that feeds IN-braid to the braid tensioner.

Element movements. Using the Developer UI, element IN and OUT movement commands are executed literally as requested. But using the Operational UI, all IN element movements are followed by an OUT. This is to remove any backlash – and it means that when going IN, there is a small over-travel IN first to compensate for the final OUT. Further, should the IN phase take you near to parked, then the element will indeed Park first – and then come OUT to the demanded position. All this arithmetic is totally transparent to you, as the user, but the result might give you a fright if you weren't expecting it.

Go mode, Jump mode. This choice of operating modes is pre-set by you to determine how PICaYAGI behaves immediately following any detected frequency change. In Go mode, you have to actively select from a menu to either go to the Solution or to change Solution and then go to that. More detail follows, but the critical feature is that if you don't select anything, absolutely nothing happens.

By contrast, if you elect to use Jumpmode, the moment a frequency change is detected, PICaYAGI jumps to it and immediately starts changing to the new Solution with no user permission required. Since changing to a different Solution requires potentially large movements of the elements, you don't want to do it casually by accident. Go mode is therefore definitely the mode of choice when first setting things up and while you are gaining familiarity and confidence.

X-mode. This is a simple ON / OFF switch. When X-mode is ON, the two linear resonators are toggled and the ELF stub switches to the 'wrong' band. The resultant retuning of the elements almost always results in a change of pattern to that of a bi-directional beam with, typically, a modest rise in SWR. Sometimes the pattern will slightly favour the reverse direction, sometimes the forward.

In any event, the ability to take an instant peek off the back of the beam is invaluable at times. If you are literally in the middle of a 3-way QSO, it is the only way that it can be done. On a conventional Yagi it can't be done at all. It is also useful if you are listening to a DX station and want to get a feel for the pileup behind you. For this to work effectively in real life, the changeover has to be near instantaneous. Even a few seconds of delay would be too long to be useable.

Solutions v Frequency. Your Tx frequency is measured and rounded to the nearest 20kHz within our 20 / 17 / 15 / 12 / 10m amateur band allocations. Although you could store tuning Solutions at 20kHz intervals, you don't actually need to find and store a Solution for every frequency you may ever want to use. The software will compute a Solution by interpolation from the nearest stored Solutions in the same band. So on the narrow 17m and 12m bands, two stored Solutions are enough. On 20m and 15m you will need more, depending on your mode and operating habits. On 10m you need several for full coverage, bearing in mind that our 10m allocation is wider than all the others put together.

Any Tx activity determines the current system frequency. The fastest dit or the smallest cough is enough. Frequency is measured twice and, if any change is detected, four times. A change is actioned only if all 4 counts are the same.

Frequency is measured with 2.5kHz resolution and there must be not less than a 7.5kHz change to trigger a change of the 20kHz interval. This built-in hysteresis is so that if, by chance, you are near a 20kHz boundary, since your instantaneous modulation may cause the Tx frequency to be continuously crossing the boundary, this 7.5kHz change requirement means the display does not dither. More significantly, the elements don't cycle in and out if in Jump mode.

OPERATIONAL UI

This gives you the ability to use pre-stored Solutions – but not to create them. If you press the Command switch on the Command Unit, PICaYAGI sends you a short menu sequence of CW characters via your Rx. You make a choice by pressing the Command Switch during or immediately after the character you want has been sent.

The other way to bring up the menu is to transmit after changing frequency while in Go mode. (As already explained, in Jump mode there is no user intervention and therefore no menu).

By design, the menu has only essential choices and the sequences are so simple, they are easily remembered and ingrained. This is important because it allows you to anticipate the menu sequence.

The top level CW menu sequence is G 3 X J P and it works as follows:

G Go to the Solution for the current frequency
3 = current Solution. If selected, the other two options are sent in numerical order – 1 2 or 2 3 or 1 3. To change to a different Solution, just select it as it goes by.
X X-mode toggle ON / OFF (initialises to OFF at power on time)
J Jump mode toggle ON / OFF
P Park – all 3 elements.

That's about all there is to it. If you have absolutely zero CW skills, you will have no problem with this interface. I am the living proof. Frankly, it is not CW in the accepted sense. Rather, it is a series of predictable noises – much like a contest QSO.

DEVELOPER UI

This runs under QBASIC on your PC and it communicates with the Controller via an RS232 link. The cabling up to the antenna can be up to 100m of cheap 3-core signal cable. Mine is bundled in with some low current rotator cores. The idea is to tune up PICaYAGI on a given frequency in order to define and save a tuning Solution that can be subsequently recalled for use on (or, by interpolation, near) that frequency thereafter. A tuning Solution comprises:

- An identifying number from 1 – 3
- The lengths of the DIR, FED and REF
- DIR linear resonator ON or OFF

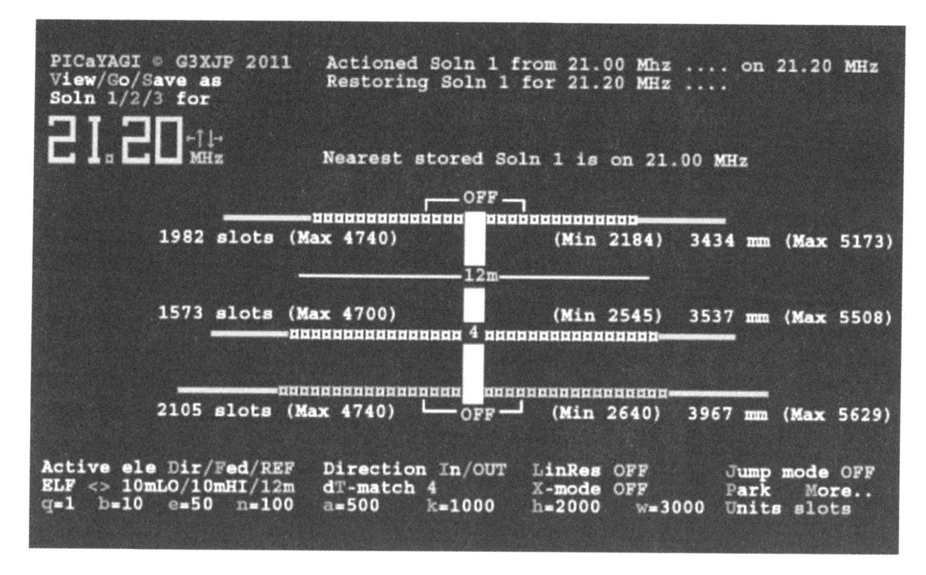

Fig 8.20: Screen snapshot of the Developer UI main menu.

- REF linear resonator ON or OFF
- ELF stub selection – 1 of 3 choices
- δT- match – 1 of 8 choices

and you can store up to 3 such tuning Solutions per frequency. You can allocate these 3 Solutions for any purpose you want. For example, mine are allocated to Best F/R, Best Gain and the third is for experimental playing.

The general process is to tune the elements like any other Yagi, but from the comfort of your shack. This includes finding the best δT match selection. As mentioned previously, this whole tuning process is highly iterative. More detail follows later. Just be grateful you are not lowering the mast for each and every adjustment!

Fig 8.20 shows the main menu to control all this. The principal feature is a scaled mimic diagram of the antenna that shows the current state of the variables. It also updates element size in real time as the element lengths actually change. Below that are the commands for altering the variables.

In the top left hand corner are controls for selecting Solution and frequency. You can manually alter the frequency within the band using the ← and → keys. Band can be changed using the ↑ and ↓ keys. At any time you can View the calculated Solution for the current frequency and then Go to it. Or you can Save the current state of affairs as Sol(utio)n 1, 2 or 3 against the current frequency.

Choosing the More ... command from the main menu gets you a sub-menu. In brief, facilities are provided to:

- Reload the PIC program in situ.
- Define the maximum and minimum lengths by element. The software will always refuse to drive the element beyond these safety stop limits.
- Zero trim. Used to manually calibrate the software and tell it that the element is at or very close to Parked.
- Timeout. A safety feature, this tells the software when to abort, should it encounter an obstruction. (My best one to date was driving the tip of an element into the uphill slope when the mast was fully lowered).
- Parking. Similar in concept to Timeout, this time-constant applies as the element slows up in the final phase of parking.

FINDING TUNING SOLUTIONS

There are several prerequisites for achieving your tuning solutions.

Infrastructure. First, you need a reliable radiated signal source. You might want to enlist the help of a fellow amateur within ground wave range (and with a horizontally polarised antenna). But I think you will try their patience if you rely on this too much to start with. Completely useless for this purpose in my experience is any sky-wave signal. The amount and rate of QSB and random shifts in polarisation are inevitably more than the differences you are looking for. That said, you can build up a feeling using a regular sked with another (constant) station. For example Harold, W4ZCB, and I have had a regular sked every week-day for years and we just 'know' how my rhombic used to perform. I had to take it down to erect and test PICaYAGI and we both came to the conclusion pretty quickly that PICaYAGI was significantly better. So this and other folklore – like how fast you break a pile-up – may have subjective merit in assessing the finished result. But they are useless as a means of tuning the Yagi to arrive at it.

Ideally, you want a signal source at least 100m away with its antenna horizontally polarised, in the clear and at least the same height as PICaYAGI. You also do not want any other conducting structures such as overhead phone or electricity lines or other HF antennas in the vicinity. This is a counsel of perfection and it is usually called an 'antenna test range'. It applies if you want to measure the results numerically with any hope of being accurate. But if you simply want to tune for the best results and are prepared to not be too disappointed by the actual numbers, these requirements can be relaxed significantly.

Ultimately, you have to live with what you have. In my experience, a Yagi tuned up under less than ideal conditions will still be about right. What you will notice is

Photo 15: My completed PICaYAGI in service on 15m.

that the rear pattern – when pragmatically tested against real propagated signals – is consistently much better than you can measure locally.

For the signal source, a DDS generator, good collection of crystals or another Tx are obvious alternatives. The output level must be small enough to not saturate your Rx and, above all, be stable in frequency and amplitude.

My best setup is described shortly, but I recognise that not everyone has that sort of space available. I have, however, also used merely a dipole in a spare bedroom at the same height as PICaYAGI and only about 20m away. The polar diagrams come out much worse but, significantly, I have been unable to find much difference in the best tuning solution.

You also need some means of plotting the pattern. I used *PolarPlot* by G4HFQ **[9]**. This relies on your Rx (with the AGC switched off) and your PC sound card being linear over about a 30dB range. My PICaSTAR Tx / Rx certainly is but my cheap generic sound card has about 1dB of compression, which shows up as about 1dB worse apparent results. If you have a known good stepped attenuator, you can verify the linearity.

For measuring the received signal, a dB-linear S-meter is very useful. Or you can use *PolarPlot* in Calibrate mode, with the Rx AGC off.

Tuning process. I suggest you start with an easy one on the 15m band. It is easy because you want both Linear Resonators switched OFF, the ELF is not used and you are nowhere near an end stop.

Starting from Parked and pointing at the signal source, the sequence I use to find Best F/R is:

1. Set δT-match to 0 (this means no series capacitor).
2. Set both LinRes to OFF. (For the bottom end of 20m you will need both ON – but not otherwise.)
3. Bring the FED out to roughly right by observing a peak in the source and / or a dip in the SWR.
4. Bring the REF out. You will see the signal dip as it passes through the right length to be a director, then the signal will come up again near the right reflector length. *Now the iteration starts.*
5. Swing the beam through 270° and settle on the largest rear lobe. Tune the REF to minimise this lobe. But not by too much without checking that some other rear lobe has not become larger.
6. Repeat 5, but this time tuning the DIR.
7. Try other δT-match values and choose the one that gives best SWR. Then try moving the FED slightly to find if either longer or shorter improves both the SWR and the F/R.
8. Loop back to 5 and iterate until you can do no better.

At this point, if it feels good, you can do a full *PolarPlot*. The process for tuning for a little more gain and a lot less F/R is the same, but with the DIR tuned instead for maximum forward gain.

The next step is critical. Although all your element lengths are now correct, they are not repeatable because, under the Developer UI, the braid backlash has not been taken out. To get it right, select each element in turn and drive it IN by 500 slots. Then bring it OUT in increasingly small increments until you get back to the same tuning point – but without doing any IN movements. Then you can validly Save the solution.

Tuning the ELF. You only have to do this once! You need to adjust the length so that the ELF works best above about 28.5MHz with the 10m HI stub selected. Use a very short 10m LO stub for contrast. Thereafter you tune the lengths of the 10m LO stub for the bottom end of 10m and the 12m stub for 12m.

A very useful trick is to make both stubs only slightly different lengths and use both of them on both bands. Switching between them tells you if the target stub

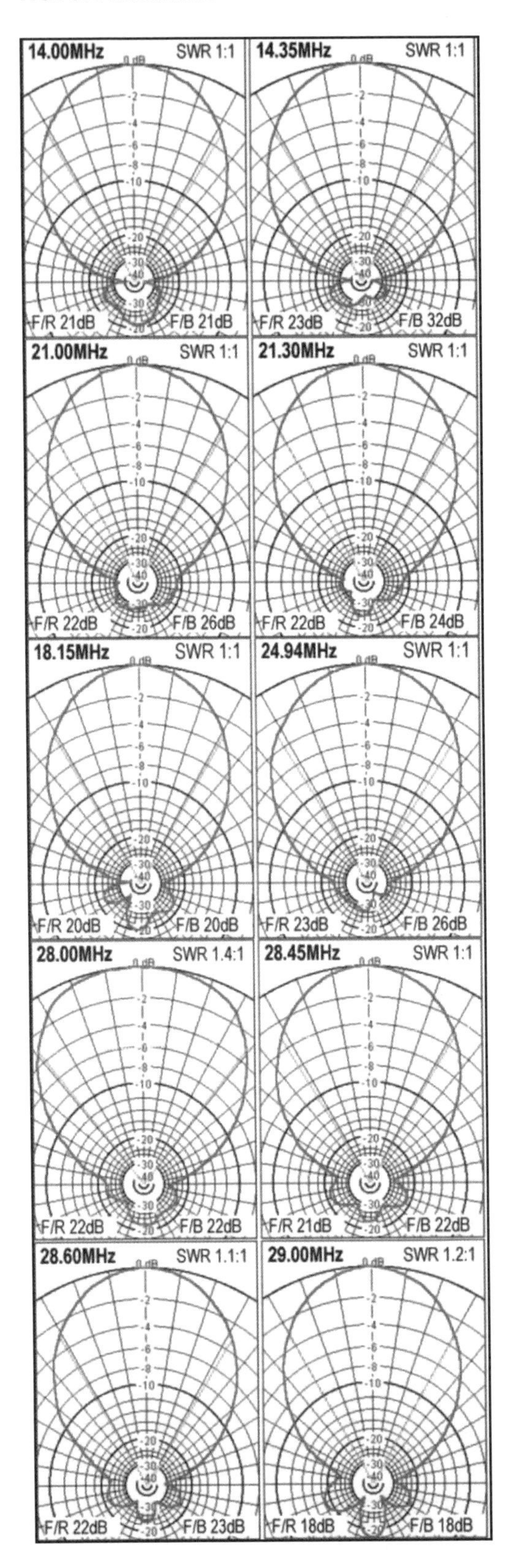

needs to be longer or shorter. I did all this with the antenna only about 3m off the ground and then refined it at full height later.

Tuning the Linear Resonators. With both Linear Resonators switched OFF, tune up on the very top end of 20m. Then switch the DIR Linear Resonator ON and shorten the DIR until you get the same result. Then shorten it a little more and trim the resonator capacitor. Repeat for the REF.

Move down the band by 30kHz and re-trim first the FED and then both resonator capacitors. Keep moving down the band until you can't get nearer to the bottom of the band without increasing the DIR and REF lengths.

MEASURED PERFORMANCE

This is the process I used to measure performance and the results I achieved. ***Measurement setup.*** For my signal source on all frequencies, I used a DDS generator. It was placed 3m up in a tree with a 12V battery and connected directly to a horizontal dipole cut for about 12m. To change frequency etc I put up an aluminium ladder, changed the settings and always replaced the ladder on the ground.

This source is about 80m away from PICaYAGI (and at the same height only because of the ground-slope). In fact, I'm firing along and slightly up a 20° slope, which is far from ideal.

My S-meter has a test mode that gives 0.01dB resolution. This is great for noting if a very small change to element length is in the right direction but adds nothing to the ultimate measurement accuracy.

Performance plots. Some representative plots are shown in **Fig 8.21**. In general, they show better than 20dB F/R and unity SWR. The performance is slightly degraded on 14.00MHz compared to the rest of 20m; you can also see it is starting to run out of adjustment at 29.00MHz.

Pattern skew. If you look closely at the patterns in **Fig 8.21**, you may notice that the rear pattern is skewed slightly. This skew could be an artefact of the antenna or equally, it could be an artefact of the measuring environment. To find out which, I flipped the beam over on its back on 10m (you wouldn't want to try this on any other band) and noted that the skew did not change sides. So I attribute it to significant quantities of fencing wire mesh and

Fig 8.21: Representative 5-band measured azimuth patterns showing F/R, F/B and SWR with the PICaYAGI at a height of 10m.

barbed wire, the ground slope and an 11kV overhead line about 100m away – but not to the antenna itself.

Forward Gain. You will note that I've made no mention so far of gain. This is because I'm not trying to sell you anything – and it is not validly measurable here or in any other domestic setting. However, here are three tentative thoughts:

1. If the rear pattern is good, then the RF must be going somewhere. Some of it is certainly going in non-useful directions in other planes. But since the rear performance is very close to that predicted by NEC modelling, it is perhaps reasonable to assume that the gain predicted by that same model won't be far out – particularly in directions where I have a clear take-off.
2. I put up a dipole for 15m that appeared not to interact. The PICaYAGI gain over that dipole into the USA was consistent with the NEC model.
3. An unique PICaYAGI capability is to radically de-tune the parasitic elements, leaving a mere FED dipole. Thus it is possible to measure gain versus that dipole-only configuration. I don't think this is entirely legitimate because both antenna modes use the same matching system. So if that were very lossy, that fact would be masked by this measurement. That said, gain measured in this way also correlates closely with the NEC model.

All this goes to prove that measurements that give the results you want are ultimately more valid than those that don't!

IN CONCLUSION

I would like to thank Ian, GM3SEK, and Harold, W4ZCB, for their suggestions for this article. Thanks also to Byron, WA4GEG, and Duke, W1ZA – and Harold, W4ZCB, in particular – for hours of patient on-air testing. It was great amateur radio – and with a great new antenna as a result!

An amateur video showing the workings of PICaYAGI has been posted to *YouTube* at **[10]**.

REFERENCES

[1] http://groups.yahoo.com/group/picaproject
[2] *RadCom*, June 2010
[3] www.arrl.org/advertising-opportunities/#Acceptance_Policy
[4] www.qsl.net/4nec2
[5] www.steppir.com/faq-frequently-asked-questions
[6] http://uk.groups.yahoo.com/group/picaproject/files/4.%20Pic-a-STAR/STAR%20documentation/
[7] www.ecfibreglasssupplies.co.uk/
[8] www.aluminiumwarehouse.co.uk
[9] www.g4hfq.co.uk/index.html
[10] www.youtube.com/watch?v=GAxJLI0vKDM

INDEX